Optimization of Spiking Neural Networks for Radar Applications

Muhammad Arsalan

Optimization of Spiking Neural Networks for Radar Applications

Springer Vieweg

Muhammad Arsalan
Technische Universität Braunschweig
Munich, Germany

ISBN 978-3-658-45317-6   ISBN 978-3-658-45318-3 (eBook)
https://doi.org/10.1007/978-3-658-45318-3

© The Editor(s) (if applicable) and The Author(s), under exclusive license to Springer Fachmedien Wiesbaden GmbH, part of Springer Nature 2024

This work is subject to copyright. All rights are solely and exclusively licensed by the Publisher, whether the whole or part of the material is concerned, specifically the rights of translation, reprinting, reuse of illustrations, recitation, broadcasting, reproduction on microfilms or in any other physical way, and transmission or information storage and retrieval, electronic adaptation, computer software, or by similar or dissimilar methodology now known or hereafter developed.
The use of general descriptive names, registered names, trademarks, service marks, etc. in this publication does not imply, even in the absence of a specific statement, that such names are exempt from the relevant protective laws and regulations and therefore free for general use.
The publisher, the authors and the editors are safe to assume that the advice and information in this book are believed to be true and accurate at the date of publication. Neither the publisher nor the authors or the editors give a warranty, expressed or implied, with respect to the material contained herein or for any errors or omissions that may have been made. The publisher remains neutral with regard to jurisdictional claims in published maps and institutional affiliations.

Planung/Lektorat: Carina Reibold
This Springer Vieweg imprint is published by the registered company Springer Fachmedien Wiesbaden GmbH, part of Springer Nature.
The registered company address is: Abraham-Lincoln-Str. 46, 65189 Wiesbaden, Germany

If disposing of this product, please recycle the paper.

Dedicated to my beloved family …

Acknowledgment

All thanks and gratitude goes to Allah for all the blessings HE has bestowed upon us. HE has given us everything we have, and often we forget about the bounties we enjoy. I would like to pledge my humble regards firstly to Allah Almighty, Who conferred on me the determination and strength to remain focused and cohesive during this project. All respects for our dear Holy Prophet Muhammad (PBUH), who enlightened our minds to recognize our Creator and thyself as the last Prophet of Allah (SWT) and a great benefactor of mankind.

I would like to express my deepest gratitude to my supervisor, Prof. Dr. Vadim Issakov (Technische Universität Braunschweig, Germany) for his exceptional guidance and support throughout my Ph.D. journey. Vadim's unwavering dedication, extensive knowledge, and ready availability have been invaluable assets to my research journey. His constant encouragement and insightful perspectives have been a driving force, inspiring me to pursue excellence in every aspect of my work. Even during demanding periods, Vadim's accessibility and willingness to provide constructive feedback have played a crucial role in meeting publication deadlines successfully. Moreover, his friendly and approachable nature has fostered a collaborative and nurturing environment within our team. I feel incredibly fortunate to have had such an exceptional mentor like Vadim throughout this academic endeavor. The wealth of wisdom he has generously shared has not only shaped me as a researcher but has also contributed significantly to my personal growth. I will always cherish the profound impact he has had on both my professional and individual journey. I extend my heartfelt gratitude to Prof. Dr.-Ing. Dr.-Ing. habil. Robert Weigel (Friedrich-Alexander-Universität Erlangen-Nürnberg, Germany) for serving as my second reviewer and examiner. His role has been pivotal in ensuring the completion of the thesis.

I am deeply thankful to the managers, Holger Schmidth and Kay Bierzynski, from Infineon Technologies AG, Germany. Their persistent support, trust, and belief in my capabilities have played a vital role in my Ph.D. journey. I am truly appreciative of the opportunities they have given me and the freedom they allowed me to explore and develop my research ideas. Their patience, guidance, and valuable feedback have been instrumental in shaping both my management skills and research approach. Their mentorship has been invaluable, and I am grateful for their contributions to my academic and professional growth.

My colleague, Avik Santra, from Infineon Technologies AG, Germany, deserves special mention for his intense support and invaluable guidance throughout my Ph.D. Avik has been a significant mentor in my academic journey, and it was through his support that I was able to join Infineon as a student. His belief in my abilities and decision to facilitate my entry into the organization not only opened up new possibilities for collaboration but also introduced me to the fascinating world of Artificial Intelligence and radar technology.

Avik's continuous encouragement, guidance, and generous sharing of his expertise have undoubtedly influenced the trajectory of my research and academic pursuit. His profound knowledge in this field, coupled with his dedication to nurturing my personal growth, has truly transformed and bolstered my professional development. I am immensely grateful for his mentorship, which has been instrumental in shaping my academic and career journey.

Gratitude extends to my friends and coauthors, as well as my colleagues at Infineon, including Moamen El-Masry, Gianfranco Mauro, Dr. Hendrik M. Lehmann, and Dr. Jakob Valtl. Their presence and support have truly made my Ph.D. journey an unforgettable and fulfilling experience. Moamen, your unwavering support and invaluable friendship have made this journey truly remarkable. Your kindness, dedication, and laughter have been a source of strength, making every success even more enjoyable. I'm deeply grateful for our shared experiences and the knowledge we have gained together. Gianco, your sense of humor, memes, and light-hearted jokes have infused our office environment with delight, bringing laughter and joy to our discussions and collaborations, making the research process much more enjoyable. Your positive energy and camaraderie have been invaluable, and I am grateful for the genuine friendship we share. Hendrik, I want to express my gratitude for being my walking partner and providing

Acknowledgment

much-needed breaks from the intensity of our research work. Our casual conversations outside the academic realm have allowed me to unwind and recharge, and your companionship has been a source of strength throughout this journey. Jakob, your invaluable knowledge and insights into the German administration have been deeply appreciated. Your role as my German translator has greatly facilitated my research and interactions within the German-speaking community. Your patience and willingness to assist have been instrumental, and I'm truly grateful for your support throughout this journey.

I am deeply grateful to my students, Zheng Tao (Infineon Technologies AG, Germany), Zechen Wang (Infineon Technologies AG, Germany), Kamran Umar (Sungrow Deutschland GmbH, Germany) and Linyan Yang (ETH Zürich, Switzerland), for their invaluable contributions to my research. Their dedication and hard work have played a vital role in overcoming challenges during my Ph.D. Their efforts have significantly advanced my research in various aspects. Without their assistance, completing this thesis would have been impossible. I am fortunate to have had such talented and diligent students accompany me on this academic journey. Your exceptional efforts and dedication are truly appreciated.

My heartfelt gratitude extends to my collaborators, Prof. Vladimir Vlassov, Sana Imtiaz, Zainab Abbas, Davide Di Matteo, and Ali Yousaf from KTH Royal Institute of Technology, Stockholm, Sweden. Prof. Vladimir Vlassov's invaluable guidance and expertise shaped my Ph.D. journey profoundly. I am genuinely thankful for his steadfast support and the priceless insights he has generously shared, which have greatly enriched my work. A special and heartfelt thank you to you, Dr. Sana Imtiaz (Epidemic Sound, Sweden). Your exceptional management and leadership skills have been a constant source of inspiration for me. Your guidance during the ANDANTE project significantly shaped my growth. Our collaboration yielded remarkable outcomes in uncharted academic territories. Your commitment to excellence profoundly influenced our research quality. Dr. Zainab Abbas (Mentimeter, Sweden), my sincere thanks go out to you for your invaluable collaborations and your remarkable ability to infuse joy into the research environment. Your contributions to our collaborative projects have been truly exceptional and has significantly enhanced the quality of our work. Davide Di Matteo (Frontiers, Netherlands) and Ali Yousaf thank you for your invaluable work during your master's thesis. Working with you has been a pleasure, and I've thoroughly enjoyed our collaboration. Learning from your insights and discoveries has significantly contributed to my research and enriched my Ph.D. journey.

I extend my heartfelt gratitude to my friends in Munich: Kashif Javed (Senior Consultant, Germany), Muhammad Atif, Muhammad Usman, and Hazra Souvik from Infineon Technologies AG, Germany. Kashif, I am grateful for your company during dinners and coffee outings, exploring various restaurants and engaging in conversations beyond the realm of PhD topics, such as current affairs and politics—a much-needed break from academic endeavors. To Atif, Usman, and Souvik, I want to express how much I appreciated our coffee breaks and the insightful conversations we shared during office hours. These moments were not only refreshing but also provided a welcome break from the intensity of our work. Your camaraderie and engaging discussions not only refreshed our minds but also uplifted our spirits, making our workdays more enjoyable and productive.

I extend my heartfelt gratitude to my esteemed friends from Belgium: Dr. Ramzan Tabasum (Senior Director, Visp, Switzerland), Nizabat Khan (Senior Service Delivery Manager, ST Engineering iDirect, Belgium), and Dr. Hafeez Muhammad Chaudhary (Senior Research Engineer, Royal Military Academy, Belgium), for generously sharing their expertise during our discussions. Being junior to all of you, I truly felt guided and supported, as if I had elder brothers by my side. Your presence has enriched me with invaluable learning experiences, profoundly shaping my communication skills and research methodologies. Our engaging conversations on politics, societal challenges, and collaborative efforts towards a better society were thoroughly enjoyed. I must acknowledge that this period has been one of the most enriching in my life.

I would like to express my deepest gratitude to my supervisor, Asifullah Khan (PIEAS, Pakistan), for his pivotal role in shaping my academic journey during my bachelor's degree. I am immensely appreciative of his significant contribution in igniting my passion for research. Asifullah's guidance, mentorship, and infectious enthusiasm for scientific inquiry were crucial in nurturing my curiosity and fostering a profound appreciation for the world of research. His belief in my potential and dedication to my academic growth inspired me to pursue further studies and embark on this Ph.D. journey. I am forever indebted to Prof. Asifullah for instilling in me the "virus of research," cultivating a relentless pursuit of knowledge, and a commitment to pushing the boundaries of discovery. The impact he had on my academic development has been immeasurable, and I attribute much of my success to the strong foundation he laid during our time together.

I extend my deepest appreciation to my friends from Pakistan, namely Tariq Habib Afridi (Phd Student, Kyung Hee University, South Korea), Umm-e- Habiba (Phd Student, University of Stuttgart, Germany), Mian Amir Iqbal, Sajid Javed

Acknowledgment

and Zavvar Bin Tariq. Tariq and Habiba, I am sincerely grateful for the enlightening group discussions we shared, delving into research topics despite our diverse fields of expertise. Your insights were invaluable and contributed significantly to my knowledge base. The exchange of ideas and expertise was not only enlightening but also instrumental in shaping my approach to research, offering fresh angles and thought paradigms that have deeply enriched my academic journey. Tariq, Amir, Sajid and Zavvar my heartfelt gratitude goes out to each of you for the exquisite gatherings we enjoyed in Pakistan whenever I took a much-needed break from my rigorous research. Those moments were like a breath of fresh air, reminiscent of our college days filled with non-stop laughter, serving as a comforting remedy to alleviate the stress that often accompanies doctoral studies. Your friendship and these cherished memories have been a source of strength throughout my academic journey.

I want to sincerely thank my parents and family for their unwavering support and encouragement throughout my academic journey. Their love, sacrifices, and belief in my abilities have been the driving force behind my accomplishments. In particular, I am deeply grateful to my mother, who, despite not having had formal education herself, devoted her life to ensuring that her children received the best education possible. Her determination, resilience, and commitment to our academic success have been a constant source of inspiration to me.

I would like to extend my deepest gratitude to my brother, Mazhar Imran, for his invaluable support, guidance, and financial assistance throughout my academic journey. His unwavering belief in my abilities and his willingness to provide both advice and financial support have played a crucial role in my success. Imran's insightful advice, drawn from his own experiences and expertise, has been incredibly valuable in shaping my academic and professional pursuits. His wisdom, encouragement, and practical guidance have been a guiding light, helping me navigate through various challenges and make informed decisions. Moreover, I am immensely grateful for Imran's generous financial support. His willingness to invest in my education and research has lightened the financial burden and allowed me to fully focus on my studies and research endeavors. His belief in the significance of education and his commitment to supporting my aspirations have been a tremendous blessing in my life.

I would like to extend my heartfelt gratitude to my brother, Zeeshan Umar, for his support and for being instrumental in my journey to Germany. Over the years, we have shared countless memories and experiences that have made my time abroad truly unforgettable. Having Zeeshan by my side has provided me with comfort and reassurance, making me feel at home, even when we were far away from our family. Together, we have explored new cuisines, enjoyed cooking experiments, and embarked on exciting outings and trips. These moments of joy and laughter have played a crucial role in maintaining a sense of balance and happiness throughout my Ph.D. journey. Zeeshan's companionship has been an invaluable source of strength and encouragement, and I am truly grateful for his presence and support throughout this transformative experience in Germany.

Furthermore, I want to express my deep gratitude to Infineon Technologies AG for granting me the invaluable opportunity to collaborate and work with them during my Ph.D. journey. Being able to work alongside such a prestigious and leading semiconductor company has been an exceptional experience, enriching my research and providing valuable insights into the industry.

Last but certainly not least, this has been made possible through funding from various sources. It was supported by the Electronic Components and Systems for European Leadership Joint Undertaking under grant agreement No. 826655 (Tempo), which, in turn, received support from the European Union's Horizon 2020 research and innovation program, with partners in Belgium, France, Germany, Switzerland, and the Netherlands. Additionally, funding was provided by the German Federal Ministry of Education and Research (BMBF) within the KI-ASIC project (16ES0992K), a part of the KI-Element call. Furthermore, support was received from the ECSEL Joint Undertaking (JU) under grant agreement No. 876925 (ANDANTE), with contributions from the European Union's Horizon 2020 research and innovation programme, as well as from France, Belgium, Germany, the Netherlands, Portugal, Spain, and Switzerland.

Abstract

Edge devices play a pivotal role in enabling the growth and success of IoT applications by providing real-time processing, reduced latency, increased efficiency, improved security, and scalability. The adoption of AI on edge is expected to grow rapidly, as more and more devices become connected to the internet and the need for real-time data processing continues to increase. This trend is expected to drive innovation in the field of AI, fostering the development of cutting-edge applications and use cases.

Although traditional deep neural network (deepNet) approaches, running on AI accelerators, are the most effective for most machine learning tasks, there is a concern about their energy efficiency during inference, especially for edge devices. The energy consumption of deepNets predominantly results from the Multiply-accumulate (MAC) operations between layers, and researchers are working to reduce the energy consumption of these operations by using tiny networks, pruning approaches, and weight quantization. Furthermore, the physical size of the AI accelerators may not fit well for the edge devices.

In recent times, Spiking Neural Networks (SNNs) have gained popularity for their energy efficiency, predominantly owing to the accessibility of resources needed to develop effective hardware for SNNs. Unlike deepNets, SNNs use spike timing to communicate information, including latencies and spike rates. SNNs rely on sparse communication, meaning that information is only transmitted when a neuron's membrane potential attains a designated threshold. This sparsity of communication, characterized by 1-bit activity, substantially diminishes the volume of data exchanged among nodes. Moreover, as nodes merely combine the received spikes, adders replace MAC arrays, resulting in a substantial reduction in required computations.

Gesture recognition offers a seamless and instinctual interface between humans and computers, surpassing traditional input modalities like keyboards and mouse. Radar technology exhibits potential for portable gesture recognition systems, capable of capturing and processing subtle finger motions to extract meaningful information from the gestures made. In this thesis, we initially propose two novel approaches to gesture recognition using a 60-GHz frequency-modulated continuous wave radar (FMCW) radar and SNNs operating on range-Doppler maps. Our study provides experimental evidence of the efficacy of SNNs, attaining a recognition rate of 98.5% and 97.5% in identifying four gestures such as swipe up-down, down-up, left-right, and finger rub. These results are on par with the performance of deepNets counterparts.

We then further extend and propose a novel SNN-based approach that can classify up to 8 hand gestures. The proposed system employs a preprocessing step involving 2D-FFT across fast-time and slow-time dimensions to generate range-Doppler maps. These maps are subsequently processed to extract range spectrograms, Doppler spectrograms, and angle spectrograms. The resulting spectrograms are inputted into an SNN for the gesture classification task. Experimental results demonstrate that the SNN, with a minimal number of neurons, achieves recognition accuracies approaching 99.50% for eight dynamic gestures, comparable to deepNets counterparts. Additionally, the proposed SNN model has a compact size of 75 kB, making it memory efficient compared to the state-of-the-art.

Traditional radar data processing methods heavily rely on computationally expensive Fourier transforms, leading to significant energy consumption. In this study, we present a novel SNN-based system that operates solely on the raw Analog-to-Digital Converter (ADC) data of the target, thereby eliminating the need for time-consuming slow-time and fast-time Fourier transforms (FFTs). Additionally, our proposed SNN architecture mimics the slow-time FFT, achieving a remarkable processing speed of 112 ms, which surpasses the performance of existing methods by more than a factor of 2. Through extensive experimentation, we demonstrate that despite the simplified approach, our proposed implementation attains a gesture recognition accuracy of 98.1% for four distinct gestures, which is on par with existing approaches.

We further extend this work by proposing a novel SNN radar system that utilizes the raw ADC data of a target, eliminating the need for slow-time and fast-time FFTs. The proposed architecture incorporates and mimics slow-time FFT and fast-time FFT directly into the SNN, eliminating the requirement for FFT accelerators. This design specifically caters to gesture sensing applications. Through experimental evaluation, the system achieves an impressive gesture classification accuracy of 98.7% for eight different gestures. This performance is comparable to conventional approaches while offering advantages such as reduced complexity, lower power consumption, and faster computations.

Finally, in the context of gesture sensing, we present an FPGA-based demonstration of SNN gesture sensing. Our demonstration involves controlling a lamp using SNN executed on an FPGA. The lamp is capable of executing four distinct functions corresponding to four specific gestures, namely, turning on, turning off, changing color, and adjusting brightness. Through our implementation on Xilinx FPGAs, we have achieved a commendable level of accuracy, reaching 98.5%.

Air-writing systems empower users to write characters or words in unrestricted space using finger or hand movements. This thesis introduces various novel radar-based airwriting systems for gesture input, offering improved accuracy, energy efficiency, and reduced storage requirements compared to existing approaches. Traditional air-writing systems based on radar utilize a network of radars, commonly three or more, to determine the position of the hand target through trilateration and trace the path of the written character. However, the actual deployment of these systems is hindered by the possibility for miss detection of the finger or hand target by all three radars. Additionally, the placement of multiple radars is not always feasible or cost-effective. In addition, these solutions fail to fully leverage the potential of deepNets, which have the capability to learn features implicitly. In contrast, we first propose a system in this thesis that employs a network of sparse radars, specifically fewer than three, to sense the local hand trajectory. We propose an architecture called 1D DCNN-LSTM-1D transposed DCNN, which combines 1D deep convolutional, long short-term memory, and transposed convolutional layers. This architecture aims to reconstruct and categorize the drawn character using solely range information obtained from each radar. By conducting experiments with one and two 60-GHz millimeter-wave radar sensors, we substantiate the effectiveness of the proposed air-writing solution. The attained accuracies are 97.33% ± 2.67% and 90.33% ± 4.44% for the solutions based on two radars and one radar, respectively, surpassing alternative deepNets architectures.

In addition, we propose a novel method that leverages one or two radars to sense the localized hand trajectory. To jointly extract features and capture the temporal attributes, we employ a 1D temporal convolutional network (TCN), enabling the identification of drawn characters from the local target trajectory.

The outcomes of the experiment reveal that the end-to-end solution achieves impressive mean accuracy rates of 99.11% and 91.33% for two radar and one radar-based approaches, correspondingly, surpassing alternative deepNets methods typically employed in this field. Additionally, the proposed system enables uninterrupted word formation through continuous character writing.

In order to tackle the issue of energy consumption, we additionally propose a novel air-writing system that prioritizes high energy efficiency by employing SNNs. The trajectory of the air-written character ascertained through accurate range assessments and trilateration using a network of radars, is identified and categorized by an SNN. The system we propose attains a similar degree of accuracy in classification (98.6%) when compared to state-of-the-art deepNets.

Lastly, in the context of air-writing we propose a novel radar-based air-writing system using a SNN and a Genetic Algorithm (GA) for optimizing the parameters of the Spiking Legendre Memory Unit (SLMU). The proposed approach works on a single radar. This solution offers similar accuracy levels to deep learning equivalents, with improved energy efficiency and smaller memory footprints. Experimental results demonstrate an accuracy of (98.53%) for two radars and (95.37%) for a single radar, with estimated energy consumption of 2.04 μJ, outperforming deepNets counterparts that consume energy in the order of mJ.

As machine learning advances rapidly, the healthcare industry is transitioning from a focus on treatment to prevention. Smart healthcare relies on large-scale health datasets for machine learning-based services. However, protecting individuals' privacy is crucial when sharing sensitive personal information. To address this, synthetic datasets generated by generative models emerge as a promising solution for privacy-preserving data sharing. Among these models, Generative Adversarial Networks (GANs) have showcased impressive results in recent times. Nevertheless, smart healthcare data presents unique challenges, including data size, various data formats and distributions.

In this thesis, we propose combining a generative adversarial network and privacypreserving mechanisms to generate realistic and confidential smart healthcare datasets. Our approach generates synthetic data samples that closely resemble the original dataset while preserving privacy through differential privacy mechanisms, even when noise is added during the learning process. We validate our approach using a real-world Fitbit dataset, where we illustrate its ability to generate synthetic and confidential datasets of exceptional quality. Moreover, our method ensures the preservation of the statistical characteristics inherent in the original dataset.

Finally, we propose a novel energy-efficient and privacy-preserving forecasting system utilizing SNNs on real-world health data streams. We compare it to a state-of-the-art system utilizing LSTM-based prediction models. In our evaluation, we observe that SNNs trade-off accuracy, resulting in a 2.2× higher error rate. However, this trade-off offers advantages such as a more compact model with 19% fewer parameters, a 77% reduction in memory consumption, and a 43% decrease in training time. Moreover, our model demonstrates a substantial reduction in energy consumption, estimated as 3.36 µJ, in comparison to conventional ANNs. To enhance privacy guarantees, ε-differential privacy into our models based on federated learning. By setting a differential privacy parameter of $\varepsilon = 0.1$, our experiments reveal a mere 25% rise in the recorded average error (Root Mean Square Error). This outcome effectively demonstrates the efficacy of the privacy protection measures in place.

Zusammenfassung

Edge-Geräte spielen eine entscheidende Rolle für das Wachstum und den Erfolg von IoT-Anwendungen, da sie Echtzeitverarbeitung, geringere Latenzzeiten, höchste Effizienz und verbesserte Sicherheit und Skalierbarkeit, im Gegensatz zu zentralisierten Lösungen bieten. Es wird erwartet, dass der Einsatz von KI im Edge-Bereich schnell zunimmt, da dies eine exzellente Möglichkeit der Datenverarbeitung großer Datenmengen darstellt.

Traditionelle Deep Neural Network (DeepNet)-Ansätze, die auf KI-Beschleunigern laufen, eignen sich gut für die meisten Aufgaben des maschinellen Lernens. Deren Einsatz auf Edge-Geräten ist aufgrund ihrer Energieeffizienz jedoch nicht zwangsläufig die beste Lösung, insbesondere beim Einsatz auf batteriebetriebenen Geräten. Der Energieverbrauch von Deep-Nets resultiert hauptsächlich aus den Multiplikations-Akkumulationen (MAC). Hauptaugenmerk liegt auf der Reduktion des Energieverbrauchs dieser Operationen durch die Verwendung kleinerer Netze, Pruning-Ansätze und Gewichtungsquantisierung. Dennoch sind bereits existierenden Ansätze nicht auf die Anforderungen von Edge Anwendungen, sondern größere Netze zugeschnitten.

Ein KI-Ansatz der zur Einsparung von Energie erforscht wird sind Spiking Neural Network (SNN). Diese Netze sind effizienter als DNNs, da SNNs die Information in Spike-Timings sowie deren Latenzen und Raten kodiert. Des Weiteren ist die SNN Kommunikation spärlich, was bedeutet, dass Kommunikation nur dann stattfindet, wenn das Membranpotenzial eines Neurons einen bestimmten Schwellenwert erreicht. Diese reduzierte Kommunikation verringert das Volumen der zwischen den Knoten ausgetauschten Daten und damit den Energieverbrauch erheblich. Da die Neuronen lediglich die empfangenen Spikes kombinieren, reduzieren sich die MACs, was zu einer deutlichen Verringerung der erforderlichen Berechnungen führt.

In dieser Arbeit werden die Fähigkeiten von SNNs in der Radardatenverarbeitung eingehend untersucht und optimierte Algorithmen entwickelt, wobei der Schwerpunkt auf zwei spezifischen Anwendungsfällen liegt deren Anwendungsgebiet die Gestenerkennung und Air-writing ist. Um die vielfältigen Einsatzmöglichkeiten der entwickelten Methoden zu belegen, wurden die Untersuchungen auf den Fitbit-Fitness-Tracker als weiteren Anwendungsfall ausgedehnt. Durch diese Erweiterung konnte die Vorhersagefähigkeit von SNNs im Zusammenhang mit der Analyse von Fitnessdaten effektiv demonstriert werden.

Contents

1 Introduction 1
  1.1 Preface 1
  1.2 Thesis Statements 2
  1.3 Objectives 2
  1.4 Contributions 4
  1.5 Publications 6
  1.6 Thesis Outlines 9

2 Background 11
  2.1 Radar Fundamentals 11
      2.1.1 Radar Equation 12
      2.1.2 Frequency Modulated Continuous Wave Radar 13
            2.1.2.1 Chirp 14
            2.1.2.2 Simplified FMCW Radar 14
            2.1.2.3 Target Range 15
            2.1.2.4 Fast Fourier Transform 17
            2.1.2.5 Range Resolution 19
            2.1.2.6 Target Velocity 20
            2.1.2.7 Vibrating Source 22
            2.1.2.8 Angle Estimation 23
  2.2 Artificial Neural Networks 25
      2.2.1 Multiple Layer Perceptron 25
      2.2.2 Convolutional Neural Network 26
            2.2.2.1 Convolutional Layer 27

		2.2.2.2	Pooling Layer	28
		2.2.2.3	Dense or Fully Connected Layer	29
		2.2.2.4	Nonlinearity Layer	29
	2.2.3	Long Short-Term Memory Networks		29
	2.2.4	Temporal Convolutional Networks		31
	2.2.5	Generative Adversarial Networks		33
		2.2.5.1	GAN Loss Function	33
		2.2.5.2	Boundary-seeking GAN	34
2.3	Spiking Neural Networks			35
	2.3.1	General Spiking Neural Network Architecture		36
	2.3.2	Leaky Integrate and Fire Model		36
	2.3.3	Encoding Schemes		38
		2.3.3.1	Rate-based Encoding	38
		2.3.3.2	Temporal Encoding	39
2.4	Neural Engineering Object (Nengo) Simulator			41
	2.4.1	NengoDL		41
2.5	Legendre Memory Unit			41
2.6	Federated Learning			42
2.7	Differential Privacy			42
	2.7.1	Laplacian Noise Addition Mechanism		43

3 **Signal Processing Chain with Spiking Neural Networks for Radar-based Gesture Sensing** ... 45

3.1	Introduction			45
	3.1.1	State-of-the-Art		46
		3.1.1.1	2D Range Doppler Images	47
		3.1.1.2	Detection Algorithm	50
		3.1.1.3	Clustering	51
		3.1.1.4	Radar Feature Images	52
		3.1.1.5	Deep Classifiers	54
	3.1.2	Limitations of State-of-the-Art		54
	3.1.3	Proposed Solutions		55
3.2	Hardware			57
	3.2.1	System Parameters		59
	3.2.2	Signal Model		60
3.3	Dataset and Experimental Setup			61
3.4	Gesture Sensing Using Range-Doppler Images			62
	3.4.1	Signal Processing Chain		62
		3.4.1.1	Target Detection	63

		3.4.1.2 Gesture Classification	63
	3.4.2	Architecture & Learning	64
		3.4.2.1 Architecture	64
		3.4.2.2 Model Testing	64
	3.4.3	Results & Discussion	65
		3.4.3.1 Dataset	65
		3.4.3.2 Classification Results	66
		3.4.3.3 Discussion	66
	3.4.4	Conclusion	68
3.5	Optimized Gesture Sensing Using Range-Doppler Images		68
	3.5.1	Signal Processing Chain	68
	3.5.2	Architecture & Learning	69
		3.5.2.1 Loss Function	69
		3.5.2.2 Model Testing	70
	3.5.3	Results & Discussion	70
		3.5.3.1 Dataset	70
		3.5.3.2 Classification Results	70
		3.5.3.3 Discussion	70
	3.5.4	Conclusion	73
3.6	Gesture Sensing Using Range-Doppler and Angle Images		74
	3.6.1	Signal Processing Chain	74
		3.6.1.1 Preprocessing	74
		3.6.1.2 Coherent Pulse Integration	74
		3.6.1.3 Moving Target Indication Filtering	74
		3.6.1.4 Target Detection	76
		3.6.1.5 Range-Doppler Image	76
		3.6.1.6 Angle of Arrival Estimation	76
		3.6.1.7 Data Generation For the Model	76
	3.6.2	Architecture & Learning	77
		3.6.2.1 Architecture	78
		3.6.2.2 Loss Function	78
		3.6.2.3 Learning Schedule and Weight Initialization	79
		3.6.2.4 Model Testing	79
		3.6.2.5 Model Hyperparameters	79
	3.6.3	Results & Discussion	81
		3.6.3.1 Dataset	81
		3.6.3.2 Results	81
		3.6.3.3 Discussion	82

		3.6.4	Conclusion	87
3.7	Gesture Recognition System Using Raw ADC Data			88
	3.7.1	Signal Processing Chain		88
		3.7.1.1	Moving Target Indication Filtering	88
	3.7.2	Architecture & Learning		89
	3.7.3	Results & Discussion		90
		3.7.3.1	Dataset	90
		3.7.3.2	Results	90
		3.7.3.3	Discussion	91
	3.7.4	Conclusion		93
3.8	Mimicking Fourier Transforms with Spiking Neural Networks			93
	3.8.1	Signal Processing Chain		93
		3.8.1.1	Raw Data	94
		3.8.1.2	Moving Target Indication Filtering	94
	3.8.2	Architecture & Learning		95
		3.8.2.1	Training	97
		3.8.2.2	Testing	98
	3.8.3	Results & Discussion		98
		3.8.3.1	Dataset	98
		3.8.3.2	Results	98
		3.8.3.3	Discussion	98
	3.8.4	Conclusion		109
3.9	Spiking Neural Networks-based Gesture Sensing on FPGA			110
	3.9.1	Building a Model in Nengo		111
	3.9.2	Model Transformation to Lower Level Code		112
	3.9.3	Implementation on the FPGA		113
	3.9.4	Conclusion		113
4	**Radar-based Air-writing for Embedded Devices**			**115**
4.1	Introduction			115
	4.1.1	State-of-the-Art		116
		4.1.1.1	Range Estimates	118
		4.1.1.2	Target Localization with Trilateration	118
		4.1.1.3	Trajectory Smoothening Filters	119
		4.1.1.4	Trajectory Reconstruction and Classification	120
	4.1.2	Limitations of State-of-the-Art		120

		4.1.3	Proposed Solutions	120
4.2	Hardware			122
		4.2.1	System Parameters	124
4.3	Dataset and Experimental Setup			124
4.4	Air-writing With Sparse Network of Radars			126
		4.4.1	Signal Processing Chain	126
		4.4.2	Proposed System	127
		4.4.3	Architecture & Learning	131
			4.4.3.1 Architecture	131
			4.4.3.2 Loss Function	133
			4.4.3.3 Weight Initialization & Learning Schedule	133
		4.4.4	Results & Discussion	134
			4.4.4.1 Dataset	134
			4.4.4.2 Reconstruction Results	134
			4.4.4.3 Classification Results	137
			4.4.4.4 Discussion	137
		4.4.5	Conclusion	140
4.5	Radar Trajectory-based Air-writing Recognition			141
		4.5.1	Signal Processing Chain	141
		4.5.2	Proposed Solution	141
		4.5.3	Architecture & Learning	143
			4.5.3.1 Loss Function	144
		4.5.4	Results & Discussion	144
			4.5.4.1 Dataset	144
			4.5.4.2 Classification	144
			4.5.4.3 Discussion	145
		4.5.5	Conclusion	149
4.6	Radar-based Air-Writing Using Spiking Neural Networks			149
		4.6.1	Signal Processing Chain	149
		4.6.2	Architecture & Learning	151
		4.6.3	Results & Discussion	151
			4.6.3.1 Dataset	151
			4.6.3.2 Data Augmentation	152
			4.6.3.3 Hyperparameters	152
			4.6.3.4 Performance Evaluation	153
			4.6.3.5 Estimated Energy Consumption	154
		4.6.4	Conclusion	156

4.7 Radar-based Air-Writing System Using Spiking Legendre
       Memory Unit 156
       4.7.1 Signal Processing Chain 157
       4.7.2 Architecture & Learning 157
             4.7.2.1 Architecture 158
             4.7.2.2 Model Training and Testing 159
       4.7.3 Results & Discussion 159
             4.7.3.1 Dataset 159
             4.7.3.2 Classification Performance 160
             4.7.3.3 Discussion 160
             4.7.3.4 Performance Evaluation 164
       4.7.4 Conclusion 164

5 **Time Series Forecasting of Healthcare Data** 165
  5.1 Introduction 165
      5.1.1 State-of-the-Art 166
      5.1.2 Limitations of the State-of-the-Art 169
      5.1.3 Proposed Approaches 169
  5.2 Hardware 171
  5.3 Dataset 172
  5.4 Synthetic and Private Smart Health Care Data Generation
       Using GANs 173
       5.4.1 Data Processing Pipeline 173
             5.4.1.1 Data Collection and Imputation 174
       5.4.2 Architecture & Learning 175
             5.4.2.1 Generator Network 175
             5.4.2.2 Discriminator Network 176
             5.4.2.3 Learning Rule 177
       5.4.3 Results & Discussion 177
             5.4.3.1 Experiments 177
       5.4.4 Conclusion 182
  5.5 Time-Series Forecasting on User Health Data Streams 182
       5.5.1 Data Processing Pipeline 182
             5.5.1.1 The Clustering Mechanism 183
             5.5.1.2 Data Model for Training SNNs 184
             5.5.1.3 Clustered Federated Learning
                     and Differential Privacy 185
       5.5.2 Architecture & Learning 185
             5.5.2.1 Loss Function 186

		5.5.2.2	Learning Schedule	186
		5.5.2.3	Model Testing	186
		5.5.2.4	Model Hyperparameters	187
	5.5.3	Evaluation Methodology		187
		5.5.3.1	Dataset	187
		5.5.3.2	Performance Metrics	188
	5.5.4	Results & Discussion		189
		5.5.4.1	Baseline Model	189
		5.5.4.2	Clustered Federated Learning	190
		5.5.4.3	Differentially Private Learning	191
		5.5.4.4	Differentially Private Federated Learning	192
		5.5.4.5	Energy Estimation of SNN Model	193
	5.5.5	Conclusion		194
6	**Conclusion and Future Directions**			195
	6.1	Summary of the Results		196
	6.2	Future Work		199
Bibliography				201

Abbreviations

Adam	Adaptive Moment Estimation
ADC	Analog Digital Conversion
ANN	Artificial Neural Network
AoA	Angle of Arrival
CFAR	Constant False Alarm Rate
CNN	Convolutional Neural Network
CNN-LSTM	Convolutional Neural Network Long Short-Term Memory Networks
DCNN	Deep Convolutional Neural Network
DFT	Discrete Fourier Transform
DP	Differential Privacy
FFT	Fast Fourier Transform
FL	Federated Learning
FMCW	Frequency Modulated Continuous Wave
GANs	Generative Adversarial Networks
GA	Genetic Algorithm
GRU	Gated Recurrent Unit
HH	Hodgkin-Huxley model
HMI	Human Machine Interface
HMM	Hidden Markov Model
IF	Intermediate Frequency
IM	Izhikevich Model
ISI	Inter-Spike Interval
LIF	Leaky Integrate-and-Fire

LMU	Legendre Memory Units
LSTM	Long Short Term Memory
MIMO	Multiple Input Multiple Output
ML	Machine Learning
MLP	Multi-Layer Perceptron
MSE	Minimum Mean Square Error
MTI	Moving Target Indicator
Nengo	Neural Engineering Object
NEF	Neural Engineering Framework
PCA	Principal Component Analysis
RAI	Range Angle Image
RCS	Radar Cross Section
RDI	Range Doppler Image
RELU	Rectified Exponential Linear Unit
RNN	Recurrent Neural Network
RX	Receiving Antenna
SNN	Spiking Neural Network
SNR	Signal to Noise Ratio
STDP	Spike timing-dependent plasticity
t-SNE	t-Distributed Stochastic Neighbor Embedding algorithm
TCN	Temporal Convolutional Networks
TTFS	Time-to-First-Spike
TX	Transmitting Antenna

List of Figures

Figure 2.1 Schematic representation of a radar system, showing the transmission of electromagnetic waves by a transmitter antenna and the subsequent reception and processing of the reflected signal by a receiver antenna to obtain information about the target's position and speed 12

Figure 2.2 a) Representation of chirp on a plot of amplitude versus time and (b) depiction of chirp in the frequency domain 14

Figure 2.3 Simplified block diagram of an FMCW radar with a single transmission (Tx) and receiving (Rx) antenna. The Tx transmits a synthesized chirp that is reflected by the target and received by the Rx. The mixer combines the transmitted and reflected signals to produce the Intermediate Frequency (IF) signal 15

Figure 2.4 Illustration of the signal processing in a radar system when a static object is located at a distance R. (a) The transmitted chirp signal (green line) is reflected back from object to the receiver with a time delay of t_d, resulting in a delayed and shifted reflected signal (blue line). The mixer output is obtained by subtracting the reflected signal from the transmitted signal. (b) The resulting difference signal (orange line) represents the Intermediate Frequency (IF) signal, which contains only one frequency due to the presence of a single object in front of the radar. The constant difference between the transmitted and reflected signals, St_d, where S is the slope of the chirp, is also shown 17

Figure 2.5 Illustration of the use of FFT in FMCW radar to determine the range of an object. The transformation of the time domain signal into the frequency domain produces a single peak for a single object, as shown 18

Figure 2.6 Two signals with same frequency but differ by 90° phase resulted in FFT peaks at same locations 20

Figure 2.7 Illustration of the phase change induced by the reflected signal from the target in FMCW radar. The difference in the initial phases of the input signals determines the starting phase of the IF signal, which is affected by the phase change induced by the reflected signal. The phase offset of the transmitted chirp at point D relative to point A is reflected in the phase of the IF signal at point F, as shown 21

Figure 2.8 Illustration of the time-varying behavior of a vibrating source, such as a human heart, and the resulting variation in the phases of the reflected chirps in FMCW radar. The source undergoes minor displacement Δd in both directions, and a series of consecutive chirps are transmitted by a radar positioned in front of it, resulting in peaks of reflected chirps occurring at the same location but with varying phases 23

List of Figures   xxxiii

Figure 2.9   Illustration of the time-evolution of phase in FMCW radar for a vibrating source, and its relation to displacement and periodicity. By plotting the phase of the peak across time, as demonstrated in the figure, the maximum phase shift $\Delta\phi$ can be related to the maximum displacement Δd, allowing for the determination of the vibration magnitude. Additionally, the graph's periodicity is proportional to the vibrations 23

Figure 2.10 (a) Angle of arrival (AoA) determination illustrated, showing signal reflection angle. (b) Estimation process using object distance variations and phase changes in peaks, with a minimum of two Rx antennas. (c) Max field of view determined by $\frac{2\pi l \sin(\theta)}{\lambda} < \pi$, achieving largest angle at $l = \lambda/2$ 24

Figure 2.11 A simple multilayer perceptron (MLP) with two hidden layers 26

Figure 2.12 An example of convolutional neural network (CNN) illustrating the two main components—feature extraction and classification. The feature extraction component, comprising convolution layers, non-linear activation layers, and pooling layers, forms the core of the CNN. For classification tasks, a fully connected layer is typically incorporated 27

Figure 2.13 Graphical demonstration of convolutional operations on an input image using a kernel. As the kernel shifts over the image, pixel values are multiplied and summed to produce an output 28

Figure 2.14 A graphical illustration of the max pooling operation, showing a fixed-size window sliding over the feature map and selecting the maximum value within the window. The output feature map is a downsized version of the original map, with reduced spatial resolution, enabling better feature extraction and more efficient computation in deep learning models. 28

Figure 2.15 Example of an LSTM Cell 31
Figure 2.16 Data flow in temporal convolutional network 32

Figure 2.17 The conventional process of image generation using Generative Adversarial Networks (GANs). Initially, a generator produces an image based on random noise, which is then evaluated by a discriminator. The discriminator assigns probabilities to both real and synthetic images, denoting authentic and fake images respectively, with 1 and 0. Subsequently, the discriminator output is utilized in a feedback loop with genuine images, and the generator output is looped with the discriminator to enhance the quality of the generated images 33

Figure 2.18 Schematic representation of the LIF neuron model as a parallel RC circuit. The LIF neuron is modeled as an equivalent circuit consisting of a parallel combination of a resistor (g_L) and a capacitor (C_m), where the resistance represents the leak conductance and the capacitance represents the membrane capacitance. The circuit is driven by an input current (I) and generates output spikes when the membrane potential reaches the threshold voltage (V_{thres}). The parallel RC circuit model provides a simplified yet effective way to understand the basic functioning of the LIF neuron [31] 37

Figure 2.19 Illustration of the behavior of a LIF-neuron when exposed to a spike train input, generates output spikes only after the refractory period has ended [31] 38

Figure 2.20 The visualization showcases the utilization of temporal coding techniques, where a dashed line denotes the rising and falling edge of the stimulus. The parameter Δt represents the time difference between the reference point and the spike. In panel (f), the sequence of spikes is enumerated on the right-hand side for clarity [35] 40

Figure 3.1 Conventional radar processing pipeline for target detection and classification 47

Figure 3.2 (a) The steps involved in transforming raw data to range domain, (b) depicts the Doppler transformation steps, and (c) shows the background subtraction process 49

List of Figures xxxv

Figure 3.3	2D CFAR Detector	50
Figure 3.4	a) Infineon's BGT60TR13C FMCW radar chipset utilized in this study, b) and its simplified block diagram	58
Figure 3.5	The proposed system flow involves feeding raw radar data into a signal processing block, where fundamental tasks like range FFT and Doppler FFT are performed. The resulting data is then passed to a gesture classification block that employs an SNN to classify the gestures	60
Figure 3.6	Examples of the various types of gestures in the dateset, namely: a) Up-down gesture, b) Down-up gesture, c) Left-right gesture, d) Right-left gesture, e) Rubbing gesture, f) Diagonal southeast to northwest gesture, g) Diagonal southwest to northeast gesture, h) Clapping gesture	62
Figure 3.7	Signal processing steps showing the transformation of range-Doppler image (RDI) to range-Doppler vector (RD) and then fed to the model. A RDI for each frame is created and then the vector associated with the highest value (RD vector) is extracted. The RD vector is then inputted into the target classification block	63
Figure 3.8	Proposed spiking neural network architecture, consisting of an input layer with 64 × 1 nodes, a convolutional layer with a filter size of 3, 32 filters, and a stride of 2, a dense layer with 4 neurons, and a dense output layer of size 4. After the convolutional layer, the output is converted into spikes using LIF as a non-linear function. The network is trained using SoftLIF activation and a multi-class cross-entropy objective function	64
Figure 3.9	An example of a gesture along with its RD map of four gestures, selected from the dataset in Sect. 3.3. Which include moving the hand in a top-down or down-up direction, moving the hand towards the left, and rubbing two fingers	65

Figure 3.10   Firing patterns for the four different gesture samples
             are shown where the SNN starts firing for a sample
             from the correct class after a few time steps   67
Figure 3.11   The proposed SNN architecture includes an input
             layer of 64 × 1 nodes, two dense layers of 32 and 4
             neurons coupled with LIF as the activation function,
             and an output layer. During training, SoftLIF
             activation and backpropagation are used. For testing,
             SoftLIF is replaced with LIF to create the SNN
             and the training parameters are used to connect
             spiking LIF neurons   69
Figure 3.12   Figure showing, (a) the four different gestures
             in the dataset, and the RD vector's map of four
             distinct gesture samples in (b), along with their firing
             patterns generated by the proposed SNN model in (c).
             The firing patterns demonstrate that the SNN begins
             to fire for a sample from the correct class after a few
             time steps due to presenting each image for a longer
             duration, resulting in improved accuracy   72
Figure 3.13   Confusion matrix of the proposed SNN
             with the testing dataset, illustrating the confusion
             between down-up and finger rub gestures,
             and predominantly with right-left gestures due
             to similarities in their RD vector's maps   73
Figure 3.14   The proposed system flow involves the raw data
             from the radar being initially input to a signal
             processing block, which carries out essential tasks
             like range FFT and Doppler FFT. The resulting data
             is then transferred to a gesture classification block
             that utilizes an SNN for classifying the gestures   75

List of Figures

Figure 3.15 The signal processing workflow for generating input to our proposed model: (a) performing preprocessing steps and applying 1D FFT along fast time, (b) conducting 1D FFT along slow time, (c) utilizing Capon beam-former to estimate the angle of arrival, (d) collecting and extracting range-Doppler image (RDI) over frames, (e) collecting and extracting range angle image (RAI) over frames, (f) stacking the RDI and RAI, and (g) using them as input to the model in the form of range-time, velocity-time, and angle-time images 77

Figure 3.16 The proposed SNN architecture with a 14 × 14 × 3 input layer followed by a convolutional layer with 16 filters of size 3 and a stride of 2, appended with LIF, and a dense layer with 32 neurons appended with LIF. The network is trained with SoftLIF activation and a multi-class cross-entropy function, while in the testing phase, the trained network is reconstructed with LIF neurons. The weights and biases learned during training are used to connect spiking LIF neurons for testing, and the test inputs are presented multiple times or steps for accurate measurement of the spiking neuron output over time 78

Figure 3.17 The effect of presentation time on accuracy is shown, indicating that accuracy rises with an increase in the number of presentation steps until it reaches a plateau. The LIF model achieves peak accuracy with fewer steps compared to other models. To maintain consistency with state-of-the-art results for the LIF neuron, a presentation time of 25 steps was utilized for our comparisons 82

Figure 3.18　Figure shows a 2D visualization of the low-dimensional features extracted from the SNN for 8 different gestures using t-SNE. The plot demonstrates that the SNN has learned separable and discriminative features, with the 8 gesture classes forming tight clusters in just 4 dimensions of embedding space. This suggests that the SNN is capable of reliably classifying the 8 gesture classes even in low-dimensional feature space 83

Figure 3.19　An illustration of the output of the SNN model for different gesture classes. Each row represents a sample from one of the eight gesture classes, and the columns (a), (b), and (c) show the range, velocity, and angle spectrogram, respectively. The firing choice of the SNN model for each sample is displayed in column (d). Starting from top row the eight gesture classes are 0—down up, 1—up down, 2—left-right, 3—rubbing, 4—right-left, 5—diagonal southwest to northeast, 6—diagonal southeast to northwest, and 7—clapping 84

Figure 3.20　The proposed SNN architecture's initial layer mimics the behavior of the discrete Fourier transform (DFT). This layer consists of $2 \times N_S$ nodes, calculating real and imaginary values using weights determined by DFT trigonometric Eq. 3.25. The input data dimensions are $N_S \times N_F$, where N_S refers to total samples per chirp and N_F denotes total frames. Variables k and l range from 0 to $N_S - 1$. After this layer, a convolutional layer with 16 filters, size 3, and stride 2 follows. Its output undergoes LIF-based non-linear processing and enters a fully connected layer with 16 neurons also utilizing LIF for activation. The final model layer, with 4 neurons, handles classification 89

Figure 3.21 Examples of four distinct gesture samples (a) down-up, (b) up-down, (c) finger rub, and (d) left-right, along with the model's chosen firing patterns. It becomes evident that the SNN starts firing for the correct class samples within a short time, showcasing its ability to classify accurately. The enhanced accuracy stems from prolonging the image's presentation time, enabling spike integration over a more extended period 92

Figure 3.22 The proposed processing chain involves connecting the radar to a CPU. The CPU receives raw ADC data and performs signal processing tasks, including MTI and target detection. The filtered data is then forwarded to SNN hardware or software for gesture classification 94

Figure 3.23 The proposed SNN architecture emulates the characteristics of slow-time and fast-time FFTs in its initial layers, followed by the integration of a CNN and dense layer for classification purposes. Within this architecture: In block a) the input is multiplied with the real coefficients of the FFT, resulting in Y^r. In block b) the input is multiplied with the imaginary coefficients of the FFT, yielding Y^i. These operations mimic the range FFT. Similarly, block c) involves the multiplication of Y^r with the real coefficients of the FFT, leading to Y^{rr} and block d) represents the multiplication of Y^r with the imaginary coefficients of the FFT, leading to Y^{ri}. Likewise, block e) executes the multiplication of Y^i with the real coefficients of the FFT, generating Y^{ri}. Block f) represents the multiplication of Y^i with the imaginary coefficients of the FFT, resulting in Y^{ii}. Subsequently, the outputs Y^r, Y^i, Y^{rr}, Y^{ri}, Y^{ir}, and Y^{ii} are appended and fed to the convolution layer, followed by the dense and output layers 96

Figure 3.24   Examples of the firing patterns of the SNN for each gesture class. The classes consist of: a) 0—Down Up, b) 1—Up Down, c) 2—Left-Right, d) 3—Rubbing, e) 4—Right-Left, f) 5—Diagonal Southwest to Northeast, g) 6—Diagonal Southeast to Northwest, h) 7—Clapping. It is evident that after a few time steps, the SNN starts to fire in accordance with the correct class, indicating its ability to accurately recognize and classify the gestures ...   100

Figure 3.25   The confusion matrices of the models listed in Table 3.15 using the test dataset. Model_3 (our proposed model) shows good performance compared to the other models, due to the first layers mimicking the FFTs. The values in the confusion matrices are shown in percentages, and the axes correspond to the 8 classes of gestures: 0—down up, 1—up down, 2—left-right, 3—rubbing, 4—right-left, 5—diagonal southwest to northeast, 6—diagonal southeast to northwest, 7—clapping 103

Figure 3.26   The t-SNE plots that depict the dimensionality reduced embedded feature clusters in varying dimensions, ranging from 2 to 64. The subfigure (e) highlights that even in lower dimensions, the SNN model can learn separable and discriminating features 105

Figure 3.27   The high-dimensional feature spaces of different layers in Model_1, Model_2, and Model_3. In a), the t-SNE visualization is shown for the input, CNN, and dense layers of Model_1. b) depicts the t-SNE for the range model (Model_2), while c) illustrates the t-SNE for the layers of the proposed method (Model_3). The legend in each sub-figure corresponds to the classes as follows: 0—down up, 1—up down, 2—left-right, 3—rubbing, 4—right-left, 5—diagonal southwest to northeast, 6—diagonal southeast to northwest, 7—clapping. It can be observed that as we move deeper into the network, the features become more discriminative, leading to improved separability among different gesture classes 106

Figure 3.28	Comparison of energy efficiency in SNN hardware with other deep learning hardware for keyword spotting. The dynamic energy cost per inference was evaluated using commercially available accelerators. The experiment showed up to 10x improvement in power efficiency for SNN hardware [113]	109
Figure 3.29	FPGA based gesture sensing demonstration using SNN to control a lamp	110
Figure 3.30	The deployment pipeline created for the SNN FPGA board	111
Figure 3.31	The optimized architecture for FPGA: Begins with an input size of ($14 \times 14 \times 3$), followed by a convolutional layer using a (2×2) kernel and 3 filters, succeeded by a dense layer featuring 32 neurons, and concluding with a final dense layer having 4 neurons. LIF activation is applied to all layers except the final one	112
Figure 3.32	Schematic diagram of the FPGA implementation	113
Figure 4.1	Categorization of air-writing techniques based on sensor used, with vision-based and non-vision-based sensors as the two main categories	116
Figure 4.2	Overview of an air-writing system using millimeter-wave FMCW radars. a) Steps for range estimation: FFT spectral analysis, coherent integration, and MTI filtering. b) Trilateration for precise hand marker estimation in 3D coordinates, followed by motion tracking with a smoothening filter. Hand motion is transformed to a 2D image for DCNN gesture classification or tracked marker coordinates are input to LSTM for classification [127]	117
Figure 4.3	(a) Shows the Infineon BGT60TR24B 60-GHz radar sensor, while (b) presents a simplified block diagram of the chipset	123
Figure 4.4	Figure depicting the two data collection scenarios: (a) the radars positioned to create a vertical virtual board, and (b) the radars positioned to create a horizontal board [127]	125

Figure 4.5　A few examples of 2D reconstructed trajectories from the dataset, illustrating the diversity in orientation, shape, and size of the trajectories, which adds complexity to the learning process 126

Figure 4.6　Comparison between the conventional and proposed solutions for air-writing. (a) The conventional pipeline uses trilateration with three radars, $\alpha\beta$ tracking filter, and track history functionalities, with classification accomplished using 2D-DCNN or LSTM. (b) The proposed pipeline employs a 1D-DCNN for feature extraction, followed by LSTM for temporal modeling and 1D transposed DCNN for trajectory reconstruction, while classification is accomplished via a fully-connected softmax 130

Figure 4.7　Proposed 1D DCNN-LSTM-1D transposed DCNN network architecture where the architecture takes an input array of 200×1 (or 200×2 for two radars), consisting of a 1D CNN layer with 128 filters, followed by an LSTM layer with 100 filters. A dense layer with softmax activation is appended to the LSTM output, which is then fed into 10 1D CNN transpose layers. The architecture includes 32 deconvolutional filters of size 7×1 in the first three transpose layers, followed by 64 filters of the same size in the next three layers. The last three transpose layers each contain 128 filters of size 7×1. The final layer has 2 deconvolutional filters of size 7×1. The ReLU activation function is used for all convolutional and transposed layers. Dropout has been applied to the last layer of 1D CNN transpose layers to prevent overfitting 132

Figure 4.8　Comparison of reconstructed trajectories using one or two radars. Blue trajectory represents the reference trajectory, red represents the reconstructed trajectory using two radars, and green represents the reconstructed trajectory using a single radar. The system is capable of accurately reconstructing the original trajectories in both scenarios 135

Figure 4.9	Examples of misclassifications made by the system, resulting in incorrect reconstructed trajectories. Reconstructed trajectories (in orange) do not correspond to the actual characters, such as characters A, B, and G. The learned dictionary of characters is utilized to generate these reconstructions. Misclassifications by Dense 15 correspond to incorrect trajectory reconstruction by the 1D transposed DCNN. The misclassification is likely caused by an incorrect projection of the input trajectory feature into the wrong cluster by the learned LSTM	136
Figure 4.10	Visualization of 2D feature representations for single and two radar configurations using t-SNE. Each class is represented by a different color, with a total of 15 clusters depicted. However, misprojections are observed for characters A and G, which are erroneously placed in the cluster of character D. Similarly, clusters 2 and I are grouped together. These misclassifications lead to errors in reconstructions by the 1D transposed DCNN, as shown in (a). Despite these challenges, the two radar configuration demonstrates improved clustering, shown in (b), with only a few misclassified samples projected into a different character's cluster. This indicates the effectiveness of utilizing two radars for enhanced feature representation and reduced misclassifications	139

Figure 4.11　Comparison between the conventional pipeline shown in (a) and proposed solution shown in (b). The conventional pipeline involves trilateration using 3 radars, $\alpha\beta$ tracking filter, and track history functionalities. The proposed solution replaces these with 1D TCN for temporal modeling of local trajectories. The TCN extracts features through temporal modeling for each character, and classification is performed using fully-connected softmax. In contrast, the classification in the conventional pipeline is accomplished through 2D-DCNN or LSTM using global trajectory $\phi(t)$. The local trajectories coordinates are represented by $\rho_1(t)$, $\rho_2(t)$ and $\rho_3(t)$ of the three radars respectively 143

Figure 4.12　Comparison of range changes over frames for characters and numerals from two radars, represented by blue and red lines in column (a). The global coordinates of the characters and numerals are also presented in column (b), obtained by using range data from three radars and trilateration to reconstruct the global trajectory. In the case of a single radar, only the blue curve reading is fed to the system, while in the case of two radars both the blue and red curves readings are fed to the system 147

Figure 4.13　Local trajectories for two words "BAD" and "FACE" are presented in (a) and (b), respectively, with the separation of each word indicated. The black dotted horizontal line represents the minimum range threshold established when the target is within the field of view (FoV) or virtual board. The simple thresholding technique, evaluated on over 50 words with character lengths varying from 2 to 4, achieved 100% segmentation accuracy. For the single radar configuration, only the blue curve is fed to the system, while for the two radar configuration, both the red and blue curves are fed to the system 148

Figure 4.14 Coloumn (a) displays the signal processing steps involved in obtaining range information from FMCW radar. Spectral analysis is performed along each chirp to obtain time delay estimates, which are then combined using pulse integration. An MTI filter is applied to filter out stationary objects, and target detection and selection are performed through thresholding. Once range information from three radars is available, trilateration (shown in (b)) is used to determine the global 3D coordinates with respect to a reference. (c) shows the architecture of the proposed SNN. The network takes in 64 × 64 2D images and comprises of two convolutional layers with 32 and 64 kernels, respectively and the output layer has 15 nodes. SoftLIF is used as an activation function for all the layers in training. During testing, the SoftLIF neurons are replaced with spiking LIF, and the trained biases and weights are used to construct the network. The classification is achieved using a softmax classifier, which provides the classification probabilities. The presentation time, which represents the time duration for capturing the spiking output, is set to 25 150

Figure 4.15 The firing patterns of the proposed SNN model for six different characters and numerals (2, H, I, A, 5, and C). The model starts firing for the correct class after a few time steps, and the accuracy of the model improves with longer integration of spikes over time 155

Figure 4.16 Energy efficiency comparison of different hardware for keyword spotting application [113]. The dynamic energy cost per inference is measured using commercially available accelerators. SNN hardware demonstrates up to a 10x improvement in power efficiency compared to other deep learning hardware 156

Figure 4.17 The flow of the Genetic Algorithm (GA) and Spiking Legendre Memory Unit (SLMU) hybrid model shows the optimization process using GA to find the optimal parameters for training the SLMU, which is evaluated based on the accuracy of the SLMU. The GA operates in a population-based manner where new individuals are created through crossover and mutation, and the fitness function is used to select the best individuals for each generation. The process continues until the GA outputs the best individual that yields the highest accuracy for the SLMU 158

Figure 4.18 The prflow, involves training the model with rate-based (non-spiking) neurons using a conventional backpropagation technique. During inference, the non-spiking neurons are replaced with spiking neurons in a new model. The weights and biases learned during training are then transferred to the test model 159

Figure 4.19 Evolution of accuracy with respect to generations for single radar. The graph shows an improvement in accuracy as the number of generations increases. The optimal parameters generated by GA at the end of the generations lead to a high accuracy rate 162

Figure 4.20 Examples of the local trajectories of characters and numerals, where the system receives only the green curve reading when using one radar and both the green and red curve readings when using two radars 163

Figure 5.1 Visualization of communication rounds in the Federated Learning (FL) process with clustering, assumed to be grouped into k clusters 167

Figure 5.2 Noisy learning: clustered FL using streaming k-means. The baseline model is a traditional FL setup (not shown separately) 167

Figure 5.3 Noisy data: a) Baseline FL 168
Figure 5.4 Noisy data: b) Clustered FL using streaming k-means 168
Figure 5.5 Fitbit Charge 2 HR wristwatch tracker [148] 171

Figure 5.6   The proposed system's data processing pipeline involves aggregating and correcting time-series data, followed by transforming the data to ensure privacy and utility. This includes removing nonessential information, normalizing features, and adding controlled noise for privacy. The model is trained using a novel BGAN architecture to generate synthetic data samples, considering different privacy settings. Finally, the data is inverse-transformed, restoring original ranges, adding missing information, and preserving privacy through noise addition based on the chosen privacy level 173

Figure 5.7   (a) The architecture of the generator network, which takes an input signal of size 15 × 1 and passes it through two dense layers with 64 and 32 neurons, respectively. A Leaky ReLU activation function with a rate of 0.2 is applied after each layer, and the final dense layer is the output layer, which uses the *tanh* activation function. (b) The architecture of the discriminator network, which receives a 15 × 1 signal as input and consists of two dense layers with 512 and 256 neurons, respectively. Both layers use Leaky ReLU activation with a rate of 0.2, and the last layer is a dense layer with a single output that applies the *sigmoid* activation function 176

Figure 5.8   The stability of the GAN model is shown. The top subplot illustrates the discriminator loss for real samples (blue), the discriminator loss for generated fake samples (orange), and the generator loss for generated fake samples (green). The losses exhibit instability early in the training phase but become stable between epochs 420 to 600, with slightly increased variance. The discriminator loss for both real and fake samples is around 0.5, while the generator loss ranges between 0.5 and 1.0, indicating that the model is expected to produce plausible samples during this period 178

Figure 5.9 The histograms illustrate the distribution of calories burned per day (kcal) for a specific group: Belgium males with a resting heart rate (RHR) of `RHR=70-75bpm` . Figure (d) represents the distribution of the original differentially private (DP) data samples with noisy input, while Figure (e) displays the distribution of the DP data generated by the BGAN. The resemblance between the two distributions is supported by a high p-value of 0.97 obtained from the Kolmogorov-Smirnov (KS) test conducted on the original noisy calorie distribution and the synthetically generated noisy calorie distribution. These findings validate the effectiveness of the proposed BGAN model in generating distributions of differentially private samples 181

Figure 5.10 The health forecasting system's pipeline involves clustering and prediction components with differential privacy (DP). The clustering uses streaming k-means to group users with similar eating patterns and pattern matching to find similarities. FL is employed to train distinct models for user clusters from clustering, adding Laplacian noise via a noise addition mechanism for DP. Predictions are made locally on clean data 183

Figure 5.11 The clustering process groups users with similar eating patterns based on centroid IDs for breakfast, lunch, and dinner from three users. Eating patterns are defined by Hamming distance from these clusters. Users with the closest Hamming distance are grouped. The forecasting model is trained per group using Federated Learning (FL) and a central generic model 184

Figure 5.12 The proposed SNN architecture involves training the model with non-spiking neurons using backpropagation. Subsequently, the trained model is reconstructed with spiking neurons, while utilizing the weights and biases obtained from the pre-trained model 185

List of Tables

Table 3.1	Radar parameters employed in our experimental setup	59
Table 3.2	Comparative analysis of classification accuracy: proposed SNN model versus other existing models	66
Table 3.3	The hyperparameters used for configuring the proposed SNN	68
Table 3.4	Comparative analysis of classification accuracy: proposed SNN model versus other existing models	70
Table 3.5	The hyperparameters used for the proposed SNN	71
Table 3.6	The hyperparameters utilized for the SNN model	80
Table 3.7	The neuron parameters employed in the proposed SNN	80
Table 3.8	The classification accuracy attained by the proposed SNN model in comparison to the existing models	81
Table 3.9	Comparing the classification accuracy of the top-performing models to achieve improved accuracy	86
Table 3.10	Comparing classification accuracy: proposed SNN model versus other SNN and conventional ANN models	91
Table 3.11	Comparison of the classification accuracy of the proposed models with other existing models, including RVA (range over time images, velocity over time images, and angle over time images) and ROT (range over time image) models	99
Table 3.12	The hyperparameters employed during the training of the proposed SNN model	101

xlix

List of Tables

Table 3.13	The parameters utilized for the LIF neuron in the proposed SNN	101
Table 3.14	The parameters utilized for Adam optimization and the weight initialization of the dense and convolutional layers	102
Table 3.15	A comparative analysis of classifications based on various layers, utilizing identical parameters as depicted in Tables 3.12, 3.13, and 3.14	107
Table 3.16	The impact of post-training quantization on accuracy	108
Table 4.1	The radar system parameters employed within this study	124
Table 4.2	The average mean squared error (L_{MSE}) for the reconstruction	134
Table 4.3	Comparison of classification accuracy: proposed SNN model versus other baseline models	137
Table 4.4	[Comparsion of the classification accuracy achieved by the proposed SN model with other baseline models	145
Table 4.5	The hyperparameters used for training the proposed SNN model	152
Table 4.6	Comparison of classification accuracy between the proposed solution and state-of-the-art methods	153
Table 4.7	The classification accuracy achieved by the proposed solution in comparison to the performance of state-of-the-art methods	160
Table 4.8	Optimal parameters for SLMU obtained using the Genetic Algorithm	161
Table 5.1	Recorded features of the Fitbit dataset	172
Table 5.2	Dataset features with ranges (aggregated per day)	175
Table 5.3	Example data samples from the population in Belgium with hidden attributes of age and gender	179
Table 5.4	The hyperparameters used for training the proposed SNN	187
Table 5.5	Value range of the Augmented Fitbit dataset	188
Table 5.6	Comparison of prediction accuracy between the SNN model and the LSTM baseline model, using the units specified in Table 5.5	189
Table 5.7	Comparison of the number of parameters and model size (in bytes) between the SNN and LSTM models	190

Table 5.8	Comparison of accuracy in clustered FL models utilizing SNNs. The table illustrates the change in error relative to the baseline model trained on the entire dataset	190
Table 5.9	Evaluation results after introducing noise to the training data to achieve differential privacy (DP) in the baseline model. The table presents the average change in error compared to the baseline model trained on the complete dataset	191
Table 5.10	Comparison of outcomes achieved by incorporating noise to the training data to achieve differential privacy in the clustering configuration with $\varepsilon = 0.1$. The average change in error is presented in relation to the baseline model trained on the complete dataset, employing differentially private learning (Sect. 5.5.4.3). Increases in the average change in error reflect a decrease in accuracy	193

Introduction

1.1 Preface

Edge computing is a computing model that involves relocating applications, computational data, and services from cloud servers to the network's edge. This paradigm shift has opened up exciting possibilities for Internet of Things (IoT) applications. By incorporating machine learning, connected objects can gain intelligence and effectively handle enormous amounts of data. However, despite the progress made, a significant proportion of data processing still takes place in the cloud. This approach presents numerous challenges, including infrastructure costs, reliability, security, speed, and energy consumption.

As IoT devices evolve, the deployment of machine learning and deep learning models at the edge has become possible. This empowers local real-time decision-making, streamlined preprocessing, and privacy-enhancing applications. However, traditional deep learning models pose challenges in scalability due to their high energy consumption, memory requirements, and computational demands. To enable the deployment of these models in resource-constrained environments, optimization techniques such as quantization and pruning have been developed, which are applied either during training or post-training of the neural network. Although these techniques provide a partial solution, substantial engineering efforts are still necessary to develop models that comply with hardware limitations.

An alternative solution for energy efficiency is the use of Spiking Neural Networks (SNNs) which are novel machine learning models that emulate the structure and function of biological neurons in the brain. SNNs operate by encoding information through the timing and frequency of electrical impulses, or "spikes", which are transmitted across synapses between neurons. This coding style is believed to contribute to the computational efficiency of SNNs. As a result, SNNs have emerged as a promising technology for reducing energy consumption in machine learning applications.

In this thesis, we aim to assess the viability of SNNs in the context of edge devices. To achieve this goal, we have chosen two radar-based use cases, gesture sensing and air-writing, and a time series forecasting use case in the smart healthcare domain, utilizing health data obtained with the Fitbit fitness tracker.

1.2 Thesis Statements

The limitations of current AI accelerators for edge devices due to their inefficient energy consumption can be overcome by implementing spiking neural networks on neuromorphic hardware, which mimic the human nervous system's energy-efficient transmission via spikes and highly sparse, 1-bit activity representations, replacing MAC operators with adders, thus resulting in highly energy-efficient systems.

1.3 Objectives

The primary aim of this research is to develop optimized spiking neural networks for radar applications, particularly for edge devices such as gesture sensing and air-writing. Additionally, the research aims to assess the performance of these spiking neural networks in the domain of time series forecasting for health data obtained with fitness trackers.

O1) **Analysis of the state-of-the-art methods:** To evaluate the performance of SNN for gesture sensing and air-writing using range-Doppler maps as input. To investigate the potential of SNN in use cases of the gesture sensing, air-writing, and time series forecasting of health data streams and compare its performance with other machine learning/deep learning approaches. This objective is further discussed in the introductory sections of Chap. 3, Chap. 4, Chap. 5.

O2) **Simulating the radar signal processing steps on spiking neural networks:** To investigate the feasibility of simulating traditional signal processing techniques using SNNs. The study will examine the potential of SNNs in mimicking the functionality of traditional signal processing algorithms such as fast-time Fourier Transform and slow-time Fourier transform by comparing the performance of SNNs with traditional signal processing techniques. This objective is further elaborated in Chap. 3 Sect. 3.7 and 3.8.

1.3 Objectives

O3) **To have an end-to-end system running on spiking neural networks:** To develop an SNN model that can simulate all the processing steps required for radar-based classification and achieve direct classification from raw data. To investigate the feasibility of using SNNs to perform data preprocessing, fast-time Fourier Transform and slow-time Fourier transform in a single framework. To explore the potential of SNNs in handling raw data without the need for preprocessing and feature extraction, and compare its performance with traditional machine learning approaches. This objective is further elaborated in Chap. 3 Sect. 3.7 and 3.8.

O4) **To demonstrate the concept on FPGA:** To explore the design and implementation of hardware-efficient SNN models on FPGA, and evaluate their performance in terms of power consumption, speed, and accuracy. To investigate the potential of FPGAs in accelerating the training and inference of SNNs, and compare the performance of FPGA-based SNNs with software-based implementations. This objective is further elaborated in Chap. 3 Sect. 3.9.

O5) **To develop embedded air-writing systems:** To develop embedded air-writing systems that have low memory footprints and a sparse network of radars to accommodate them in space-constrained environments and compare their performance with state-of-the-art. This objective is further elaborated in Chap. 5 Sect. 4.4 and 4.5.

O6) **To develop SNN-based air-writing systems:** To develop SNN-based air-writing systems that have low memory footprints and are highly energy efficient and compare their performance with state-of-the-art and the systems developed as part of O5). This objective is further elaborated in Chap. 5 Sect. 4.6 and 4.7.

O7) **To demonstrate the capabilities of SNN for time series forecasting application:** To investigate the feasibility of SNNs for time series applications. The study will examine the potential of SNNs in predicting the dietary behavior of Fitbit Users. This objective is further elaborated in Chap. 5 Sect. 5.5.

O8) **Privacy-preserving synthetic dataset generation:** To generate synthetic data for health data streams in the context of O7. To design efficient solutions for privacy-preserving data sharing and enable third parties to perform privacy-preserving data analytics, while ensuring user privacy is not compromised. This objective is further elaborated Chap. 5 Sect. 5.4.

1.4 Contributions

We conduct an empirical study on the use of SNNs for edge devices in the context of radar applications and fitness trackers. We investigate the potential of SNNs in gesture sensing, air-writing, and fitness tracking, and explore different aspects of SNN processing, including model design, optimization, and deployment. Our results demonstrate the feasibility of using SNNs for these applications and highlight their potential in improving the performance of edge-based systems. The study contributes to the development of efficient and accurate edge processing systems using SNNs, with potential applications in various fields.

This dissertation makes the following contributions toward addressing the aforementioned objectives:

C1) **Novel methods of SNN-based radar applications:** Edge devices, which are becoming increasingly prevalent in today's technology landscape, must meet a variety of requirements to be effective in their respective roles. One of the most critical requirements is energy efficiency, as edge devices are typically designed to operate in environments where access to power sources may be limited. As such, devices that can operate for longer periods of time on minimal power are highly sought after. Energy efficiency is a significant factor in determining the overall effectiveness and value of an edge device, and as such, is a key area of focus in the design and development of these technologies. Our research has demonstrated the effectiveness of SNNs in radar-based gesture sensing, air-writing, and time-series forecasting. Specifically, we have developed, as part of our objective O1, innovative architectures that can effectively replace traditional deep learning methods for edge applications of gesture sensing and air-writing, while maintaining comparable levels of performance.

C2) **Novel methods for simulating the signal processing steps of radar with SNN:** When it comes to radar applications, the signal processing stages tend to consume the majority of the energy, particularly during computationally expensive FFT operations. These operations require substantial processing power and can be computationally intensive, resulting in higher energy consumption overall. As part of O2, we have successfully shown that SNNs can be utilized to emulate the signal processing steps involved in traditional radar signal processing pipelines. Our approach involved proposing new architectures capable of emulating the behavior of both fast-time and slow-time Fast Fourier Transform (FFT) processes. Thus, we demonstrate the potential of SNNs in replacing traditional signal processing methods in radar technology,

1.4 Contributions

potentially leading to more energy-efficient and cost-effective systems in the future.

C3) **Novel methods for classification directly from raw data:** As highlighted in C2, the conventional radar signal preprocessing steps, including fast-time and slow-time FFTs, followed by Constant False Alarm Rate (CFAR), account for most of the energy consumption. We developed gesture sensing systems using only raw data, which overcomes the necessity for signal processing and denoising steps. The SNN can then learn these steps on its own, displaying the full capabilities of SNNs. This approach allows for a significant reduction in energy consumption while providing results comparable to those achieved through traditional methods.

C4) **Novel methods for air-writing:** The current air-writing method relies on using at least three radars to estimate global coordinates of trajectory motion through trilateration. However, in practical applications, this method can be unreliable due to the occlusion of fingers on one or more radars and missed detections, which can lead to missing or unreliable trajectory coordinates. Moreover, positioning a radar network for an optimal field-of-view (FoV) on a screen or AR-VR device isn't always possible due to the limited FoV of individual radars and physical constraints. To address these limitations, we have developed methods using a sparse radar network, which uses range information from each radar. This approach employs novel architectures to reconstruct global trajectory coordinates and recognize characters accurately, even with fewer than three radars. Using a sparse radar network reduces the risk of occlusion and missed detections, resulting in more reliable trajectory coordinates. Additionally, it eases radar network placement, suitable for scenarios with limited individual radar FoV or where placement is challenging.

C5) **Novel methods for air-writing based on SNN:** As mentioned in C1, C2 and C3, one of the pathways toward energy-efficient systems involves the utilization of SNN. As a part of O1, O2 and O6, we have developed novel SNNs-based air-writing systems methods that have a similar level of performance with the techniques developed in C4.

C6) **A novel approach to generating synthetic private data:** A new approach has been developed for generating synthetic private data, which can be useful for preserving privacy while sharing data. The goal is to generate realistic non-sensitive datasets from sensitive ones, which aligns with objective O8. Generative adversarial networks (GANs) [1] are currently the most advanced models for synthetic data generation, and the proposed approach leverages the combination of GANs and differential privacy mechanisms to generate a smart healthcare dataset that is both realistic and private. This method can generate synthetic data samples that closely resemble the original dataset while also ensuring differential privacy in various scenarios, including learning from noisy distributions or introducing noise to the learned distribution.

The effectiveness of the proposed approach is assessed by conducting evaluations on a real-world Fitbit dataset. The results revealed that the proposed approach successfully generated a high-quality synthetic dataset while maintaining differential privacy and preserving the statistical characteristics of the original dataset. These findings align with Objectives O8.

C7) **A novel SNN-based model for privacy-preserving time series forecasting of health data:** The use of health monitoring devices is becoming increasingly popular as people seek to take charge of their health and well-being. These devices are not only useful for tracking exercise and monitoring vital signs, but they can also provide valuable information that can be used to make important healthcare decisions. The energy consumption for such devices is a critical factor to consider. To address this issue as part of O1 and O7, we investigate the use of SNNs for time-series forecasting in healthcare applications. We introduce a novel time-series forecasting model based on SNNs that demonstrates comparable performance to state-of-the-art systems but with significantly lower energy consumption.

C8) **A real-time demo of gesture sensing:** As the SNN hardware was out of the scope of this research, to realize the concept of the SNN we have developed an FPGA-based demonstrator for gesture sensing as part of objective O4.

1.5 Publications

This thesis work led to 4 peer-reviewed journals, 13 conference papers and 1 book chapter which are enumerated below. The results of the conference papers P6, P7, P15 and P16 are not incorporated within this thesis.

Journals Papers:

P1) **M. Arsalan**, A. Santra and V. Issakov, Power-efficient gesture sensing for edge devices: mimicking fourier transforms with spiking neural networks. in Applied Intelligence, Springer Nature, 2022.

P2) **M. Arsalan**, A. Santra and V. Issakov, "Spiking Neural Network-Based Radar Gesture Recognition System Using Raw ADC Data," in IEEE Sensors Letters, vol. 6, no. 6, pp. 1–4, June 2022, Art no. 7001904, https://doi.org/10.1109/LSENS.2022.3173589.

1.5 Publications

P3) **M. Arsalan**, A. Santra and V. Issakov, "RadarSNN: A Resource Efficient Gesture Sensing System Based on mm-Wave Radar," in IEEE Transactions on Microwave Theory and Techniques, vol. 70, no. 4, pp. 2451–2461, April 2022, https://doi.org/10.1109/TMTT.2022.3148403.

P4) B. Vogginger, F. Kreutz, J. López-Randulfe, C. Liu, R. Dietrich, H. A Gonzalez, D. Scholz, N. Reeb, D. Auge, J. Hille, **M. Arsalan**, F. Mirus, C. Grassmann, A. Knoll, C. Mayr, "Automotive Radar Processing With Spiking Neural Networks: Concepts and Challenges," Frontiers in neuroscience, 2022.

Conferences Papers:

P5) **M. Arsalan**, A. Santra, V. Issakov, "Low Power Radar-based Air-Writing System using Genetic Algorithm-assisted Spiking Legendre Memory Unit", EuRAD 2023.

P6) **M. Arsalan**, A. Santra, V. Issakov, "Low Power Gesture Sensing System based on Target Range Using Spiking Neural Networks for Portable Devices," accepted in IEEE Radio & Wireless Week, 2024.

P7) M. G. Janjua, K. Kaiser, **M. Arsalan**, S. Schoenfeldt, V. Issakov, "Radar-Based Gesture Recognition Using Gaussian Mixture Variational Autoencoder," submitted to IEEE MTT-S International Conference on Microwaves for Intelligent Mobility 2024.

P8) **M. Arsalan**, T. Zheng, A. Santra and V. Issakov, "Contactless Low Power Air-Writing Based on FMCW Radar Networks Using Spiking Neural Networks," 2022 21st IEEE International Conference on Machine Learning and Applications (ICMLA), Nassau, Bahamas, 2022, pp. 931–935.

P9) **M. Arsalan**, D. Di Matteo, S. Imtiaz, Z. Abbas, V. Vlassov and V. Issakov, "Energy-Efficient Privacy-Preserving Time-Series Forecasting on User Health Data Streams," 2022 IEEE International Conference on Trust, Security and Privacy in Computing and Communications (TrustCom), Wuhan, China, 2022, pp. 541–546, https://doi.org/10.1109/TrustCom56396.2022.00080.

P10) **M. Arsalan**, A. Santra, M. Chmurski, M. El-Masry, G. Mauro and V. Issakov, "Radar-Based Gesture Recognition System using Spiking Neural Network," 2021 26th IEEE International Conference on Emerging Technologies and Factory Automation (ETFA), Vasteras, Sweden, 2021, pp. 1–5.

P11) **M. Arsalan**, M. Chmurski, A. Santra, M. El-Masry, R. Weigel and V. Issakov, "Resource Efficient Gesture Sensing Based on FMCW Radar using Spiking Neural Networks," 2021 IEEE MTT-S International Microwave Symposium (IMS), Atlanta, GA, USA, 2021, pp. 78–81, https://doi.org/10.1109/IMS19712.2021.9574994.

P12) **M. Arsalan**, A. Santra, K. Bierzynski and V. Issakov, "Air-Writing with Sparse Network of Radars using Spatio-Temporal Learning," 2020 25th International Conference on Pattern Recognition (ICPR), Milan, Italy, 2021, pp. 8877–8884, https://doi.org/10.1109/ICPR48806.2021.9413332.

P13) **M. Arsalan**, A. Santra and V. Issakov, "Radar Trajectory-based Air-Writing Recognition using Temporal Convolutional Network," 2020 19th IEEE International Conference on Machine Learning and Applications (ICMLA), Miami, FL, USA, 2020, pp. 1454–1459, https://doi.org/10.1109/ICMLA51294.2020.00225.

P14) S. Imtiaz, **M. Arsalan**, V. Vlassov and R. Sadre, "Synthetic and Private Smart Health Care Data Generation using GANs," 2021 International Conference on Computer Communications and Networks (ICCCN), Athens, Greece, 2021, pp. 1–7, https://doi.org/10.1109/ICCCN52240.2021.9522203.

P15) G. Mauro, M. Chmurski, **M. Arsalan**, M. Zubert, V. Issakov, "One-Shot Meta-learning for Radar-Based Gesture Sequences Recognition", International Conference on Artificial Neural Networks, 2021.

P16) S. Imtiaz, Z. N. Tania, H. Nazeer Chaudhry, **M. Arsalan**, R. Sadre and V. Vlassov, "Machine Learning with Reconfigurable Privacy on Resource-Limited Computing Devices," 2021 IEEE Intl Conf on Parallel & Distributed Processing with Applications, Big Data & Cloud Computing, Sustainable Computing & Communications, Social Computing & Networking (ISPA/BDCloud/SocialCom/
SustainCom), New York City, NY, USA, 2021, pp. 1592–1602, https://doi.org/10.1109/ISPA-BDCloud-SocialCom-SustainCom52081.2021.00213.

P17) S. Imtiaz, S. -F. Horchidan, Z. Abbas, **M. Arsalan**, H. N. Chaudhry and V. Vlassov, "Privacy Preserving Time-Series Forecasting of User Health Data Streams," 2020 IEEE International Conference on Big Data (Big Data), Atlanta, GA, USA, 2020, pp. 3428–3437, https://doi.org/10.1109/BigData50022.2020.9378186.

Book Chapter:

P18) A. Valentian, S. Narduzzi, **M. Arsalan** et al., "Tools and Methodologies for Training, Profiling, and Mapping a Neural Network on a Hardware Target", Intelligent Edge-Embedded Technologies for Digitising Industry, River Publishers Series in Communications and Networking, 2022.

1.6 Thesis Outlines

The subsequent sections of this thesis are structured as follows: Chap. 2 offers an overview of radar technology, deep learning/machine learning methods, spiking neural networks and privacy preservation, thereby establishing the essential foundational knowledge for this study. Chap. 3 introduces our proposed methods for gesture sensing, while Chap. 4 presents our proposed techniques for air-writing. In Chap. 5, we delve into the potential of SNN for time series forecasting of health data streams and discuss our proposed approaches. Finally, Chap. 6 concludes the thesis, providing a summary of our findings and suggesting avenues for future research.

Background

This chapter lays the groundwork by introducing the fundamental concepts of radar technology. Furthermore, it offers an overview of neural networks, including spiking neural networks, which have been utilized for classification and prediction tasks. Finally, the chapter provides background information on various technologies employed for time-series forecastings, such as Federated learning, synthetic data generation, and privacy preservation techniques for handling sensitive data.

2.1 Radar Fundamentals

The term Radar, which stands for Radio Detection and Ranging, was coined by F.R. Furth during his time at the Naval Research Laboratory. This technology relies on the use of radio waves to accurately measure an object's range, angle, and velocity in its surrounding environment.

Although the origins of radar detection can be traced back to the early days of the theory of electromagnetism, it wasn't until World War 2 that radar technology began to develop into a fully-fledged field. While German engineer Hülsmezer had conducted experiments in 1903 regarding the detection of radio waves reflected from a ship [2], it was not until Robert Page's claim in 1934 that radar technology was first developed for aircraft detection [3]. Since then, radar technology has greatly evolved and is now widely used in various applications, including air-defense systems, search and surveillance, weather avoidance, navigation and tracking, as well as high-resolution imaging and mapping.

A radar system comprises a transmitter antenna that emits electromagnetic waves, which are subsequently received by a receiver antenna after being reflected by a target, as depicted in Fig. 2.1. Afterward, the received signal is subjected to further processing to retrieve information about the target, such as its position and speed [2].

When radar waves come into contact with an object, some of the waves are absorbed, reflected, or scattered in various directions depending on the characteristics of the object. Materials with high electrical conductivity, particularly metals, reflect radar waves effectively. Furthermore, if the object is either approaching or moving away from the radar, the frequency of the received signal undergoes a change due to the Doppler effect, which allows for the estimation of the object's velocity [2].

Fig. 2.1 Schematic representation of a radar system, showing the transmission of electromagnetic waves by a transmitter antenna and the subsequent reception and processing of the reflected signal by a receiver antenna to obtain information about the target's position and speed

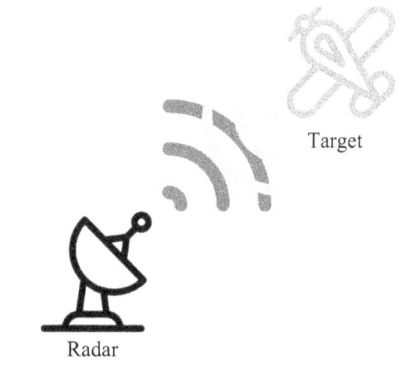

2.1.1 Radar Equation

The fundamental radar equation establishes a relation between the range or distance of a radar object and several parameters such as the transmitter and receiver properties, antenna characteristics, and target properties. Suppose a target is positioned at a distance R from a monostatic radar that transmits with a power of P_t. In this case, assuming an antenna gain of G, the power density at the target can be expressed as follows:

$$P_d = \frac{P_t G}{4\pi R^2}. \tag{2.1}$$

The radar cross section (RCS) of a target represents the area σ where the incident transmission power is concentrated on the target. If we consider a target with RCS equal to σ, which is radiating isotropically, then the power density at the radar can be calculated using the following equation [4]:

2.1 Radar Fundamentals

$$P'_d = \frac{P_d \sigma}{4\pi R^2} = \frac{P_t G_t \sigma}{(4\pi R^2)^2}. \quad (2.2)$$

Let's assume that A_e is the effective area of the radar, which can be related to the antenna gain as follows:

$$A_e = \frac{G_r \lambda^2}{4\pi}. \quad (2.3)$$

Assuming that the same antenna is used for transmitting and receiving, the received power of the radar is given by:

$$P_r = P'_d A_e = \frac{P_t G_t \sigma G_r \lambda^2}{(4\pi R^2)^2 4\pi} = \frac{P_t G^2 \sigma \lambda^2}{(4\pi)^3 R^4}. \quad (2.4)$$

If S_{min} is the minimum detectable signal, then the maximum possible distance R_{max} can be calculated as follows:

$$R_{max} = \left[\frac{P_t G^2 \sigma \lambda^2}{(4\pi)^3 S_{min}}\right]^{1/4}. \quad (2.5)$$

From the radar range equation, it is apparent that objects located near the radar transmitter generate strong reflections, whereas the strength of these reflections decreases gradually as the distance of the target from the transmitter increases. Typically, the reflected radar signal has low power, which is amplified using electronic amplifiers and advanced signal processing techniques in modern radar systems [5].

2.1.2 Frequency Modulated Continuous Wave Radar

In the initial phases of radar development, the primary emphasis was on the continuous wave (CW) systems that transmit electromagnetic waves with a fixed frequency f_0. Any reflected signal from a stationary object would have the same frequency as the transmitted signal. However, if the object is moving, the reflected signal would experience a frequency shift, denoted by f_d, due to the Doppler effect. This frequency shift contains information about the object's velocity and angular position, which can be extracted by analyzing the frequency shift. One of the main limitations of CW radar is that it does not encode any time information, which means that range information about the object cannot be obtained using CW radar systems [2].

FMCW radar, as the name implies, employs frequency-modulated continuous wave technology [6]. By modulating the carrier wave's frequency, FMCW radar can measure the distance information that was previously absent in CW radar. A chirp, a specific linear waveform where the transmission frequency changes over time, is used to introduce a timing mark [6]. The subsequent sections provide a brief overview of how FMCW radar works.

2.1.2.1 Chirp

A chirp is a sinusoidal or sine wave with a linearly increasing frequency over time [2]. Figure 2.2(a) provides an illustrative plot of the chirp's amplitude versus time. The chirp begins with a frequency of f_0 and steadily increases in frequency until it reaches a frequency of $f_c + B$, where B denotes the bandwidth of the chirp [7].

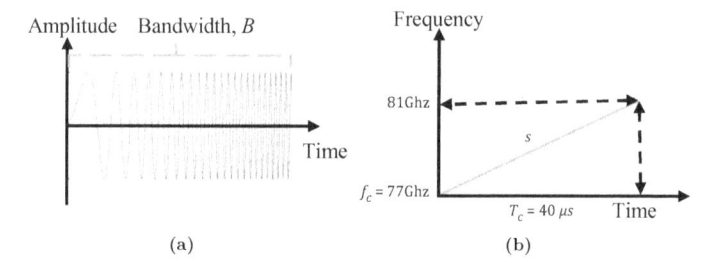

Fig. 2.2 a) Representation of chirp on a plot of amplitude versus time and (b) depiction of chirp in the frequency domain

Plotting the chirp's linearly increasing frequency will produce a straight line with a slope S in the frequency versus time plot, as illustrated in Fig. 2.2(b). The slope S indicates how rapidly the chirp ramps up in frequency. For instance, in the figure, the chirp begins at a frequency of 77 GHz, ends at 81 GHz, spans a bandwidth of 4 GHz, and has a time period T_c of 40 μsec, corresponding to a slope of 100 MHz/μsec. The bandwidth B and slope S are two crucial parameters that define the system's performance [7].

2.1.2.2 Simplified FMCW Radar

Figure 2.3 shows a simplified block diagram of a FMCW radar, consisting of a single transmission (Tx) and a single receiving (Rx) antenna. The Tx transmits a chirp generated by the synthesizer, which is then reflected by the target and received by the Rx. The transmitted and reflected signals are mixed together by the mixer, resulting in a signal called the Intermediate Frequency (IF signal). The output of the mixer has two properties [7]:

2.1 Radar Fundamentals

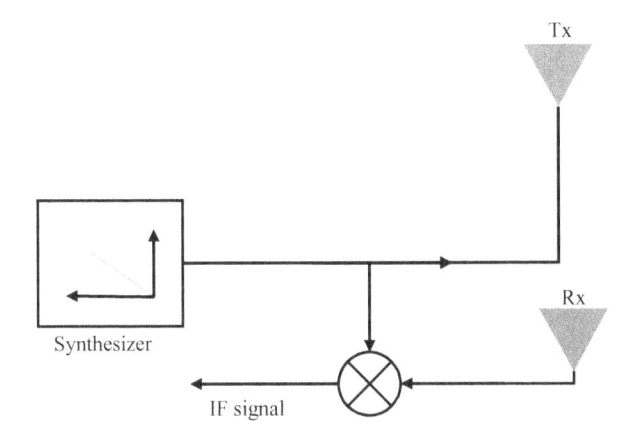

Fig. 2.3 Simplified block diagram of an FMCW radar with a single transmission (Tx) and receiving (Rx) antenna. The Tx transmits a synthesized chirp that is reflected by the target and received by the Rx. The mixer combines the transmitted and reflected signals to produce the Intermediate Frequency (IF) signal

1. The instantaneous frequency of the output is equivalent to the difference between the instantaneous frequencies of the transmitted and reflected signals.
2. The initial phase of the output is equivalent to the difference between the starting phases of the two input sinusoids.

These two properties can be represented in mathematical form as [7]:

For two input sinusoids, x_1 and x_2, with frequencies ω_1 and ω_2 and phases ϕ_1 and ϕ_2, the above properties can be mathematically expressed as:

$$x_1 = \sin(\omega_1 t + \phi_1) \; x_2 = \sin(\omega_2 t + \phi_2). \tag{2.6}$$

The mixer output x_{out} is given by:

$$x_{out} = \sin((\omega_1 - \omega_2)t + (\phi_1 - \phi_2)). \tag{2.7}$$

2.1.2.3 Target Range

A radar's transmitted chirp signal, when it encounters a static object located at a distance R, results in a reflected signal arriving at the receiver with a time delay of t_d, as shown in Fig. 2.4(a). The transmitted chirp, represented by the green line, is delayed and returned as a blue line, which is the signal reflected from the object.

The mixer output is obtained by taking the difference between the transmitted and reflected signals, as explained in the previous section. This difference signal is obtained by subtracting the blue line from the green line. The figure clearly depicts that both lines have a constant difference, i.e., St_d, where S represents the slope of the chirp. Therefore, the presence of a single object in front of a radar generates a beat or Intermediate Frequency (IF) signal, which comprises only one frequency, as illustrated in Fig. 2.4(b). This frequency is expressed as [7]:

$$f_b = St_d. \tag{2.8}$$

As t_d represents the time taken by the transmitted signal to reach the object and the reflected signal to return to the receiver, it can be expressed as the round-trip delay to the object and back, i.e.:

$$t_d = \frac{2R}{c}, \tag{2.9}$$

where c is the speed of light. Consequently, the frequency of the IF signal can be given by:

$$f_b = S\frac{2R}{c}. \tag{2.10}$$

Substituting the definition of the slope S in Equation (2.10), we get:

$$f_b = \frac{B2R}{T_c c}. \tag{2.11}$$

Thus, by knowing the values of the parameters B, T_c, and c, we can estimate the range of the object using Equation (2.11). The IF signal is considered valid only during the time when the reflected signal is received by the Rx antenna. During the digitization process, the valid IF signal is obtained by selecting the samples corresponding to the time delay t_d, and only up to the duration of the transmitted signal T_x. Typically, t_d is a small fraction of the total chirp time T_c. For instance, in a radar system with a maximum range of 300 m and a chirp time of 40 μs, the ratio of t_d to T_c is merely 5 % [7].

2.1 Radar Fundamentals

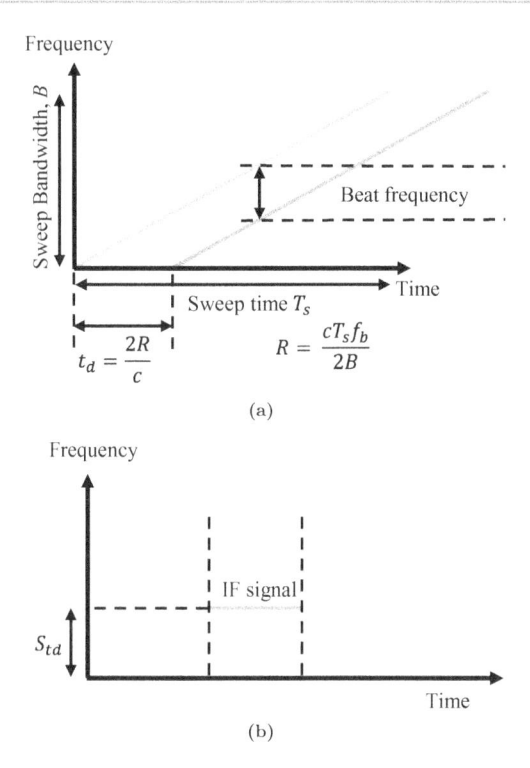

Fig. 2.4 Illustration of the signal processing in a radar system when a static object is located at a distance R. (a) The transmitted chirp signal (green line) is reflected back from object to the receiver with a time delay of t_d, resulting in a delayed and shifted reflected signal (blue line). The mixer output is obtained by subtracting the reflected signal from the transmitted signal. (b) The resulting difference signal (orange line) represents the Intermediate Frequency (IF) signal, which contains only one frequency due to the presence of a single object in front of the radar. The constant difference between the transmitted and reflected signals, St_d, where S is the slope of the chirp, is also shown

2.1.2.4 Fast Fourier Transform

In FMCW radar, the Fast Fourier Transform (FFT) is a powerful mathematical tool used to determine the range of an object. This transformation converts the time domain signal into the frequency domain, which means that a single object in front of the radar produces a single peak in the frequency domain, as illustrated in Fig. 2.5 [7].

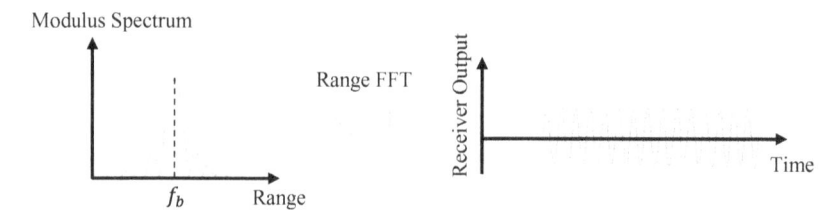

Fig. 2.5 Illustration of the use of FFT in FMCW radar to determine the range of an object. The transformation of the time domain signal into the frequency domain produces a single peak for a single object, as shown

In radar, the range FFT is used to extract the beat frequency and determine the range of the target. The frequency of the beat signal corresponds to the location of the maximum peak in the FFT spectrum, which indicates the range of the object.

Most radars digitize the IF signal for further processing. The signal is first passed through a low-pass filter and then through an analog-to-digital converter (ADC) to convert it to digital. The digitized signal is then sent to an appropriate processor, such as an FFT processor [7].

Prior to digitizing the IF signal using an ADC, it is important to determine the bandwidth of interest. This allows for the appropriate setting of the low-pass filter and the ADC sampling rate. For instance, if the maximum distance of an object is d_{max}, then the maximum frequency of the IF signal should be:

$$f_{max} = \frac{S2d_{max}}{c}. \tag{2.12}$$

Thus the bandwidth of interest for the current example should be $0 - f_{max}$. Hence, the low-pass filter should be designed with a cut-off frequency f_{max}. Likewise, it is imperative for the ADC to possess a sampling rate surpassing the cut-off frequency, denoted as f_{max}. Consequently, the maximum sampling rate of the ADC can impose a restriction on the maximum detectable distance, represented as d_{max}, by the radar system. As the maximum IF bandwidth is directly related to the slope S, and the maximum detectable distance d_{max}, both the ADC sampling rate and the IF bandwidth emerge as critical limitations in the sensor system. Typically, radars tend to use smaller S for larger d_{max} [7].

Compared to other types of correlators, the FFT correlator is much simpler when it comes to obtaining the range in FMCW radar [6]. The proportionality of range to beat frequency allows for processing of only the range bins falling within the bandwidth of interest, resulting in reduced computation and a significant decrease in the complexity of the digital processor [6].

2.1.2.5 Range Resolution
The ability of the radar to distinguish two closely spaced objects is known as range resolution. When multiple objects are present, the resulting FFT spectrum will have multiple peaks. However, if these objects are too close together, they will appear as a single peak, and the radar will display them as a single target. To distinguish between closely spaced objects in the frequency domain, the length of the signal must be increased, i.e., the observation window T_c must be increased. In general, if the frequency components of the signal are separated by more than $1/T_c$ Hz, they can be resolved. Therefore, increasing the observation window, which is proportional to increasing the bandwidth B, results in higher resolution.

To obtain an equation for range resolution, suppose two closely spaced objects are located in front of the radar. Each object will produce an IF signal of $S2R/c$, which can only be distinguished if the frequency difference between them is greater than the observation window.

If the distance between the objects is denoted by $\triangle d$, then the difference between their IF frequencies can be expressed as [7]:

$$\triangle f = \frac{S2\triangle d}{c}. \tag{2.13}$$

As the observation window is defined as T_c, it follows that:

$$\triangle f > \frac{1}{T_c}, \tag{2.14}$$

$$\frac{S2\triangle d}{c} > \frac{1}{T_c}, \tag{2.15}$$

$$\triangle d > \frac{c}{2ST_c}, \tag{2.16}$$

$$\triangle d > \frac{c}{2B}. \tag{2.17}$$

Hence, the equation for range resolution, d_r, is solely based on the bandwidth, B, swept by the chirp and is expressed as:

$$d_r = \frac{c}{2B}. \tag{2.18}$$

2.1.2.6 Target Velocity

The sensitivity of the IF signal's phase to small displacements allows the radar to precisely measure the velocity of objects, making it useful for vital sensing and vibration detection. In the frequency domain, the signal is a complex signal, where each value in the FFT is a complex number with an amplitude A and a phase θ, which can be represented as $Ae^{j\theta}$.

To demonstrate this property, signals are often represented by a phasor, where the vector length corresponds to the amplitude A and the direction corresponds to the phase θ. The phase of the peak in the FFT is equal to the initial phase of the signal in the time domain. This property can be understood by considering two time domain signals of the same frequency but different initial phases, as shown in Fig. 2.6. Although the FFT of these signals produces peaks at the same location in the FFT spectrum, the phase value of the peak for one wave compared to the other differs by 90° [7].

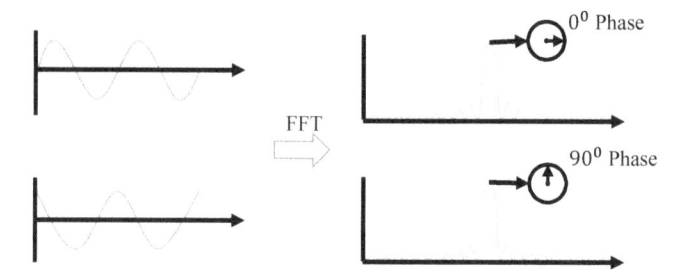

Fig. 2.6 Two signals with same frequency but differ by 90° phase resulted in FFT peaks at same locations

Let's recall from Sect. 2.1.2.2 that the difference in the initial phases of the two input signals is equal to the initial phase of the IF signal. When the reflected signal from the target arrives at the mixer after a round trip delay t_d, it induces a phase change that corresponds to the transmitted signal (the yellow line in Fig. 2.7).

Consequently, this process will affect the IF signal. The starting phase of the IF signal at point F is determined by taking the difference in phase between the Tx chirp at point D and the Rx chirp at point E, as previously discussed. At point B, the phase of the Rx chirp is the same as that of E, but the phase of the Tx chirp at point D includes an additional phase offset of $2\pi f_b \Delta t_d$ relative to the phase at point A, which is reflected at point F in the phase of the IF signal [7].

2.1 Radar Fundamentals

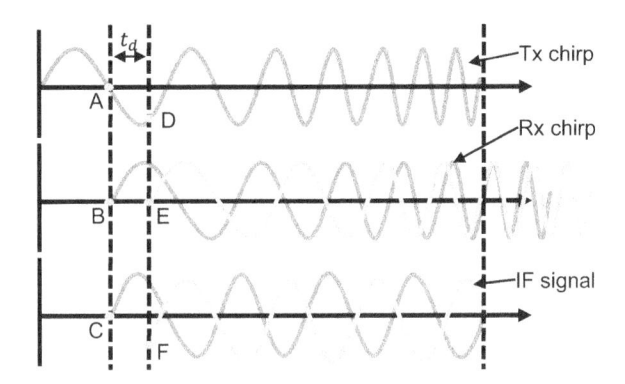

Fig. 2.7 Illustration of the phase change induced by the reflected signal from the target in FMCW radar. The difference in the initial phases of the input signals determines the starting phase of the IF signal, which is affected by the phase change induced by the reflected signal. The phase offset of the transmitted chirp at point D relative to point A is reflected in the phase of the IF signal at point F, as shown

If $\triangle\phi$ represents the change in phase, then mathematically it can be expressed as:

$$\triangle\phi = 2\pi f_c \triangle t_d, \quad (2.19)$$

where $\triangle t_d$ is:

$$\triangle t_d = \frac{2\triangle d}{c} = \frac{2\triangle d}{f_b/\lambda}. \quad (2.20)$$

By substituting the value of $\triangle t_d$ from Equation (2.20) into Equation (2.19), we obtain the change in phase as:

$$\triangle\phi = \frac{4\pi \triangle d}{\lambda}. \quad (2.21)$$

Combining all the information, the IF signal (sinusoid) produced by a single object at a distance d from the radar is given by:

$$IF = A\sin(2\pi f_b t + \phi_0), \quad (2.22)$$

where f_b is calculated using the distance d as $f_b = \frac{S2d}{c}$, and $\triangle\phi$ is calculated as $\frac{4\pi \triangle d}{\lambda}$.

The radar can measure the velocity of an object by exploiting the sensitivity of the change of phase $\triangle\phi$ to small displacements. To do so, two consecutive chirps separated by T_c are sent toward the moving object. Although the target will have moved a very small distance during the two consecutive chirps, both chirps will have a peak in the same location. Each chirp's range-FFT will exhibit a peak at the same position, but with distinct phase variations resulting from the motion of the object. Measuring this phase change allows the radar to determine the corresponding velocity of the moving object. If the object is moving with a velocity v, it will move a distance of vT_c during the time T_c, and the velocity can be calculated from the phase as [7]:

$$\Delta\phi = \frac{4\pi \Delta d}{\lambda} = \frac{4\pi v T_c}{\lambda}, \quad (2.23)$$

$$v = \frac{\lambda \Delta\phi}{4\pi T_c}. \quad (2.24)$$

2.1.2.7 Vibrating Source

In the preceding section, it was established that minor displacement can significantly affect the phase of the IF signal, to a degree comparable to the resolution of the radar, which is typically in the range of millimeters. This characteristic presents numerous possibilities for interesting applications, such as the ability to gauge the oscillation of an entity [7].

Figure 2.8 illustrates the time-varying behavior of a vibrating source, such as a human heart, as it undergoes minor displacement (Δd) in both directions. Suppose a radar is positioned in front of this vibrating source, and a series of consecutive chirps are transmitted. The reflected chirps' peaks will occur at the same location but with varying phases. By plotting the phase of the peak across time, as demonstrated in Fig. 2.9, observation can be made.

The maximum phase shift $\Delta\phi$ is related to the maximum displacement Δd, implying that determining the magnitude of $\Delta\phi$ provides the vibration's Δd. Similarly, the periodicity of the plot directly gives the period of the vibration [7].

The relationships are given by:

$$\Delta\phi = \frac{4\pi \Delta d}{\lambda}, \quad (2.25)$$

$$\Delta d = \frac{4\pi \Delta\phi}{\lambda}, \quad (2.26)$$

where λ represents the wavelength of the vibration, and the equations relate the phase shift and displacement change in terms of the wavelength.

2.1 Radar Fundamentals

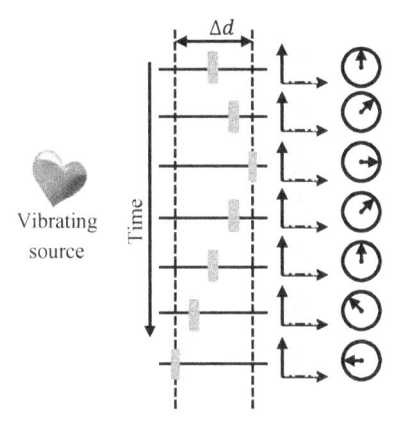

Fig. 2.8 Illustration of the time-varying behavior of a vibrating source, such as a human heart, and the resulting variation in the phases of the reflected chirps in FMCW radar. The source undergoes minor displacement Δd in both directions, and a series of consecutive chirps are transmitted by a radar positioned in front of it, resulting in peaks of reflected chirps occurring at the same location but with varying phases

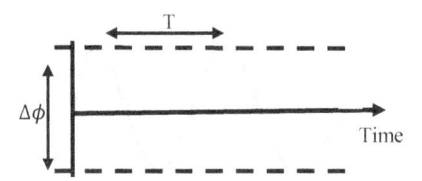

Fig. 2.9 Illustration of the time-evolution of phase in FMCW radar for a vibrating source, and its relation to displacement and periodicity. By plotting the phase of the peak across time, as demonstrated in the figure, the maximum phase shift $\Delta \phi$ can be related to the maximum displacement Δd, allowing for the determination of the vibration magnitude. Additionally, the graph's periodicity is proportional to the vibrations

2.1.2.8 Angle Estimation

The FMCW radar system possesses the capability to determine the angle at which a reflected signal arrives in relation to the horizontal plane, commonly known as the angle of arrival (AoA), as illustrated in Fig. 2.10(a). The process of estimating the angle involves leveraging the relationship between minor variations in the object's distance and the resulting phase modifications in the range-FFT or Doppler-FFT peaks. To estimate AoA, at least two Rx antennas is necessary, as depicted in Fig. 2.10(b). The accurate determination of the AoA is achieved by comparing the disparity in distance between the object and each antenna, utilizing the resulting

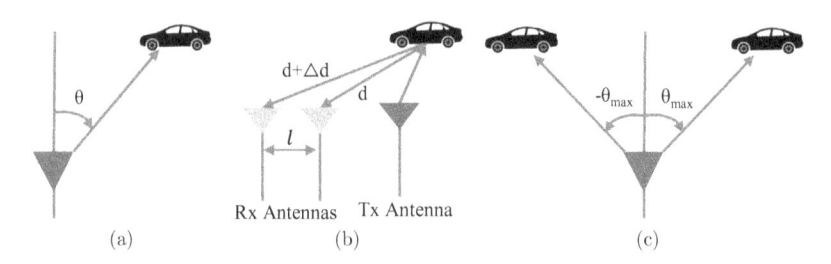

Fig. 2.10 (a) Angle of arrival (AoA) determination illustrated, showing signal reflection angle. (b) Estimation process using object distance variations and phase changes in peaks, with a minimum of two Rx antennas. (c) Max field of view determined by $\frac{2\pi l \sin(\theta)}{\lambda} < \pi$, achieving largest angle at $l = \lambda/2$

phase change in the FFT peak [7]. In this configuration, we can derive the mathematical expression for the phase change as follows:

$$\Delta \phi = \frac{2\pi \Delta d}{\lambda}. \quad (2.27)$$

Based on the assumption of a planar wavefront and basic geometry, we can establish that $\Delta d = l \sin(\theta)$, where l represents the distance between the antennas. Therefore, we can compute the angle of arrival (θ) from the measured $\Delta \phi$ using the equation:

$$\theta = \sin^{-1}\left(\frac{\lambda \Delta \phi}{2\pi l}\right). \quad (2.28)$$

It is important to note that $\Delta \phi$ is dependent on $\sin(\theta)$, introducing a non-linear relationship. However, this non-linear dependency can be approximated by a linear function when θ has a small value, i.e., $\sin(\theta) \approx \theta$. Consequently, the accuracy of the angle estimation is influenced by the AoA and tends to be more accurate when θ is small [7]. The radar's maximum angular field of view is determined by its ability to estimate the maximum AoA, as illustrated in Fig. 2.10(c). To ensure an unambiguous measurement of the angle, we require $|\Delta w| < 180°$. Using Equation (2.28), this condition corresponds to $\frac{2\pi l \sin(\theta)}{\lambda} < \pi$. The maximum field of view that can be covered by two antennas spaced apart by l is given by:

$$\theta_{\max} = \sin^{-1}\left(\frac{\lambda}{2l}\right). \quad (2.29)$$

When the spacing between the two antennas is $l = \lambda/2$, the largest angular field of view achieved is $\pm 90°$.

2.2 Artificial Neural Networks

Artificial Neural Networks (ANNs) are a group of algorithms that draw inspiration from the biological functionality of neurons present in the human brain and are considered to be one of the primary tools employed in modern-day machine learning. The first algorithm of this kind was suggested by McCulloch and Pitts in 1943 [8], where they presented the idea of logical neurons. Subsequently, Rosenblatt proposed a model in 1957 [9] that possessed the capability to learn a problem. He coined the term "perceptron" for the algorithm, which comprised a solitary neuron. The perceptron is defined as a function that accepts multiple inputs and converts them to a single output, serving as a linear classifier. Mathematically, this can be expressed as:

$$y = f(\sum_i w_i x_i + b), \tag{2.30}$$

where w_i represents the weights, b is the bias, and f denotes the activation function.

The node's output is regulated by the activation function f. The perceptron employed the Heaviside step function as its activation function [10], which resulted in a binary output of 'yes' or 'no.' Other types of activation functions, including linear classifiers and non-linear classifiers such as hyperbolic tangent, sigmoid, ReLU, parametric ReLU, Softmax, and Swish, can also be used. The main disadvantage of the perceptron, being a linear classifier, is its limitation in solving problems that are only linearly separable.

2.2.1 Multiple Layer Perceptron

To overcome the perceptron's limitations, multiple perceptrons were interconnected, resulting in the multi-layer perceptron (MLP) [11], which is capable of approximating any arbitrary continuous function. In the MLP, neurons are arranged and linked in layers, with each layer serving as an input layer to the subsequent layer, as illustrated in Fig. 2.11. The network consists of an input layer and an output layer, while the intermediate layers connecting the input and output layers are referred to as hidden layers. In this particular network, the input layer is designed to accept three inputs and map them to a single output.

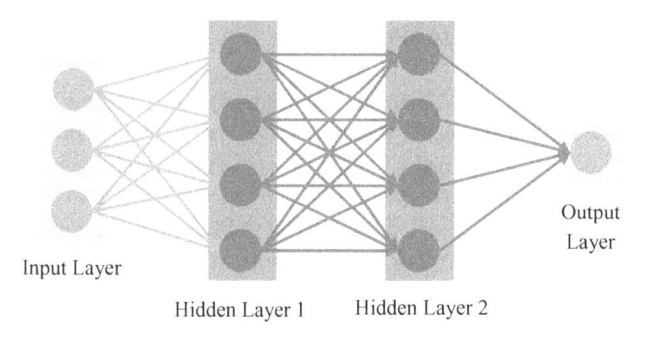

Fig. 2.11 A simple multilayer perceptron (MLP) with two hidden layers

Selecting the appropriate number of layers and the number of neurons in each layer is a crucial task in designing an MLP. Using a deep architecture with numerous layers can lead to overfitting on the training data, whereas a shallow architecture with few layers may result in underfitting. The most commonly employed activation functions in MLPs are the sigmoid [12] and hyperbolic tangent [12].

2.2.2 Convolutional Neural Network

Convolutional Neural Networks (CNNs) are a type of Neural Networks that have demonstrated remarkable efficacy in image recognition and classification. The primary distinction between CNNs and conventional ANNs is their ability to encode image characteristics into the architecture, rendering them more suitable for image-related tasks. This also lowers the number of parameters required to set up the model. The concept of CNN is based on the visual cortex and was initially proposed by Hubel and Wiesel in the 1960s [13]. However, LeCun demonstrated the first application of CNNs in 1989 for the classification of handwritten digits [14]. CNNs employ a three-dimensional organization of neurons, incorporating spatial dimensions (height and width) and depth. In contrast to standard ANNs, the neurons within each layer of a CNN are connected to a limited region of the preceding layer.

The architecture of a CNN typically consists of two main components: a feature extraction component and a classification component, as depicted in Fig. 2.12. The feature extraction component is the primary part of the CNN, which comprises a series of convolution layers, non-linear activation layers, and pooling layers. For classification tasks, a fully connected layer is usually included in the CNN. However, in certain applications where classification is not necessary, such as in autoencoders, this component is omitted.

2.2 Artificial Neural Networks

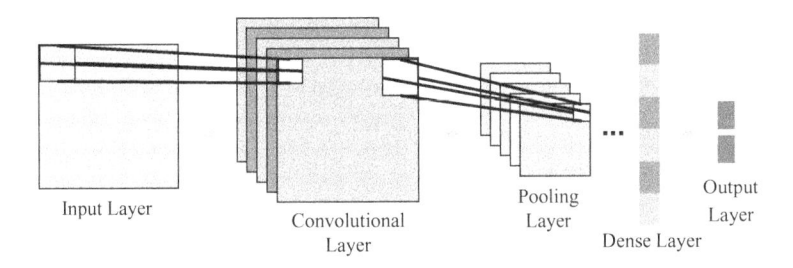

Fig. 2.12 An example of convolutional neural network (CNN) illustrating the two main components—feature extraction and classification. The feature extraction component, comprising convolution layers, non-linear activation layers, and pooling layers, forms the core of the CNN. For classification tasks, a fully connected layer is typically incorporated

2.2.2.1 Convolutional Layer

A convolution layer (Conv) is composed of several convolution filters, also known as kernels, which learn the local features in an image, creating feature maps. These features are then convolved across the image, producing an activation map matrix that represents the locations of the learned features in the input image [15]. The value of the activation map is high for a particular location in the input if the feature represented by the convolution filter is present at that location. This value is determined using the following formula:

$$h_{i,j} = \sum_{k=1}^{m} \sum_{l=1}^{m} w_{k,l} x_{i+k-1, j+l-1}, \qquad (2.31)$$

where h is the convolution output, x represents input and w represents the convolution kernel with height and width m. A visual representation of the convolutional operation is provided in Fig. 2.13.

The Conv layer is characterized by several parameters [15]:

- Number of filters/kernels, denoted as k
- Kernel dimension, denoted as f
- Stride, denoted as s
- Zero padding, denoted as p

The Conv layer accepts input data with dimensions of $w_1 \times h_1 \times d_1$, where w, h, and d represent the height, width, and depth of the input, respectively. The output data has dimensions of $w_2 \times h_2 \times d_2$, as stated in [15]:

$$w_2 = \frac{w_1 - f + 2p}{s+1}, \qquad (2.32)$$

$$h_2 = \frac{h_1 - f + 2p}{s+1}, \qquad (2.33)$$

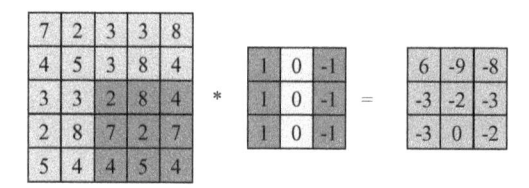

$$d_2 = k. \qquad (2.34)$$

Fig. 2.13 Graphical demonstration of convolutional operations on an input image using a kernel. As the kernel shifts over the image, pixel values are multiplied and summed to produce an output

Fig. 2.14 A graphical illustration of the max pooling operation, showing a fixed-size window sliding over the feature map and selecting the maximum value within the window. The output feature map is a downsized version of the original map, with reduced spatial resolution, enabling better feature extraction and more efficient computation in deep learning models.

2.2.2.2 Pooling Layer

Pooling downsizes feature maps in CNNs, cutting parameters and enhancing model's translation invariance [15]. Max-pooling is the most widely used pooling operation, which selects the maximum value within the kernel at each position. This can be expressed mathematically as:

$$h_{i,j} = \max\{x_{i+k-1, j+l-1} \ \forall \ 1 \le k \ge m \text{ and } 1 \le l \ge m\}. \qquad (2.35)$$

2.2 Artificial Neural Networks

The pooling layer only depends on one parameter, the stride s. It receives input with dimensions of $w_1 \times h_1 \times d_1$ and produces output with dimensions of $w_2 \times h_2 \times d_2$, where w_2, h_2, and d_2 are calculated as follows [15]:

$$w_2 = \frac{w_1 - f}{s + 1}, \qquad (2.36)$$

$$h_2 = \frac{h_1 - f}{s + 1}, \qquad (2.37)$$

$$d_2 = d_1. \qquad (2.38)$$

Figure 2.14 illustrates a graphical example of the max pooling operation.

2.2.2.3 Dense or Fully Connected Layer

Dense or fully connected layers combine features learned by different convolution kernels and produce a global representation of the input image built by the network [15]. The fully connected neurons generate distinct activation patterns depending on the features present in the input images, resulting in a compact representation that can be readily utilized by the output layer to accurately classify the image.

2.2.2.4 Nonlinearity Layer

A widely used nonlinearity function in CNNs is the Rectified Linear Unit (ReLU), which has been shown to outperform the conventional Sigmoid function [16]. The ReLU is defined as follow:

$$y(x) = \max(0, x). \qquad (2.39)$$

Compared to sigmoid functions, ReLU is more computationally efficient because it only requires computing the maximum between 0 and x, and does not involve expensive exponential operations.

2.2.3 Long Short-Term Memory Networks

MLPs and CNNs are unable to effectively handle the temporal propagation of information, which is critical for various applications such as gesture sensing and tracking. These applications necessitate a neural network capable of preserving a historical record of past events to inform future decision-making. In response to this limitation, recurrent neural networks (RNNs) [17] were introduced. RNNs utilize self-loop structures to facilitate the persistence of information across time. By leveraging these self-loops, RNNs establish connections between prior information and the current task, enabling informed decision-making based on past events. However, the widespread adoption of RNNs in the 1990s was impeded by a significant obstacle known as the vanishing gradient problem. This issue arises when network weights are repeatedly multiplied by values less than one, resulting in exponentially small values after multiple multiplications. Consequently, the gradient, a vital component for information propagation, diminishes and eventually approaches zero, hindering the ability to retain and learn from information originating from distant past instances or time points.

The following formulation describes the functioning of RNNs. When presented with a temporal input sequence denoted as $x(k) = (x_1(k), x_2(k), x_3(k))$, an RNN processes this sequence by associating it with a sequence of hidden values represented as $h(k) = (h_1(k), \cdots, h_T(k))$. Subsequently, the RNN generates a sequence of activations denoted as $a(k+1) = (a_1(k+1), \cdots, a_T(k+1))$ using the recursive equation:

$$h(k) = \sigma(W_{hx}x(k) + h(k-1)W_{hh} + b_h), \quad (2.40)$$

where the symbol σ represents a non-linear activation function, while b_h denotes the hidden bias vector. The weight matrices in the equation are denoted by W, with W_{hx} representing the input-hidden weight matrix and W_{hh} representing the hidden-hidden weight matrix.

The activation equation for recurrent units is expressed as:

$$a(k+1) = h(k)W_{ha} + b_a, \quad (2.41)$$

where W_{ha} denotes the weight matrix for hidden-to-activation connections, and b_a represents the bias vector associated with the activations.

When dealing with Recurrent Neural Networks (RNNs), the challenge of vanishing or exploding gradients arises. This challenge can be effectively addressed by employing specialized architectures such as Long Short-Term Memory (LSTM) [18] or Gated Recurrent Unit (GRU) [19]. LSTMs enhance RNNs by incorporating memory cells and gating mechanisms. These gating mechanisms utilize element-wise multiplication of input signals, enabling each memory cell to regulate its behavior. The LSTM updates its cell state based on the activation of these gates. The input provided to an LSTM is directed to various gates responsible for specific operations on the cell memory, including writing (input gate), reading (output gate), and resetting (forget gate). Through these gates, incoming signals are processed, determining whether to block or pass information based on its strength and relevance, which is determined by learned filter weights. During the learning process, these weights are adaptively adjusted to control the entry, retention, or deletion of data within the cells [20].

2.2 Artificial Neural Networks

The equations below provide the vector representation that encompasses all units within an LSTM layer:

$$\begin{cases} i(k) &= \sigma_i(W_{ai}a(k) + W_{hi}h(k-1) + W_{ci}c(k-1) + b_i), \\ f(k) &= \sigma_f(W_{af}a(k) + W_{hf}h(k-1 + W_{cf}c(k-1) + b_f), \\ c(k) &= f(k)c(k-1) + i(k)\sigma_c(W_{ac}a(k) + W_{hc}h(k-1) + b_c), \\ o(k) &= \sigma_o(W_{ao}a(k) + W_{ho}h(k-1) + W_{co}c(k) + b_o), \\ h(k) &= o(k)\sigma_h(c(k)). \end{cases} \quad (2.42)$$

The above equations describe the relationships between the input gate vector i, forget gate vector f, output gate vector o, cell activation vector c, and the hidden value vector h. The dimensions of these vectors are consistent with that of h. The non-linear functions are denoted by σ. The input to the memory cell layer at time k is represented as $x(1), x(2), \cdots, x(K)$. The computations involve weight matrices $W_{ai}, W_{hi}, W_{ci}, W_{af}, W_{hf}, W_{cf}, W_{ac}, W_{hc}, W_{ao}, W_{ho}$, and W_{co}, with their respective subscripts indicating the relationships. The bias vectors are denoted as b_i, b_f, b_c, and b_o. Figure 2.15 provides a visual representation of a single unit within an LSTM block.

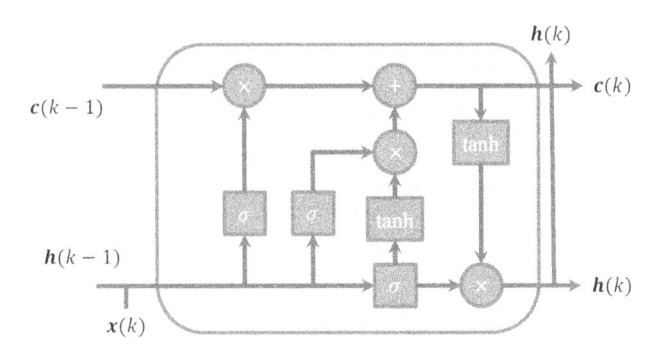

Fig. 2.15 Example of an LSTM Cell

2.2.4 Temporal Convolutional Networks

Temporal Convolutional Networks (TCN) were proposed by Lea [21] that draw inspiration from modern convolutional architectures designed for sequential data. TCN leverages dilated convolutional neural networks and have been successfully applied to various tasks, such as speech modeling [22] and recognition of human actions [23]. The TCN operates based on two fundamental concepts: dilated convolution and causal convolution. By utilizing causal convolution, the TCN ensures that information does not flow from the future to the past. In other words, it exclusively takes into account past and present inputs to predict each time step. This feature is essential in real-time prediction situations where only past and present data are available for analysis and forecasting.

As more layers are added, the receptive field of causal convolutions expands progressively, allowing them to incorporate historical information up to the network's depth. However, this poses a limitation. To overcome this constraint, TCN utilizes dilated convolutions, which elongate the convolutional kernels and amplify the receptive field. This extension enables the kernel to capture longer-term dependencies and address the issue of limited historical information. The dilation convolution is defined for an input sequence $x \in \mathbb{R}^T$ and a filter $h : \{0, \cdots, k-1\} \to \mathbb{R}$ as [21]:

$$H(x) = (x *_d h)(x) = \sum_{i=0}^{k-1} f(i) x_{s-d \cdot i}, \tag{2.43}$$

where $d = 2^\ell$ represents the the dilation factor, with ℓ as the network level, and $x_{s-d \cdot i}$ represents the past direction. By increasing the dilation factor, the top-level output can encompass a wider spectrum of inputs, thereby expanding the receptive field of a convolutional neural network. The data flow of the TCN employed in this study is depicted in Fig. 2.16.

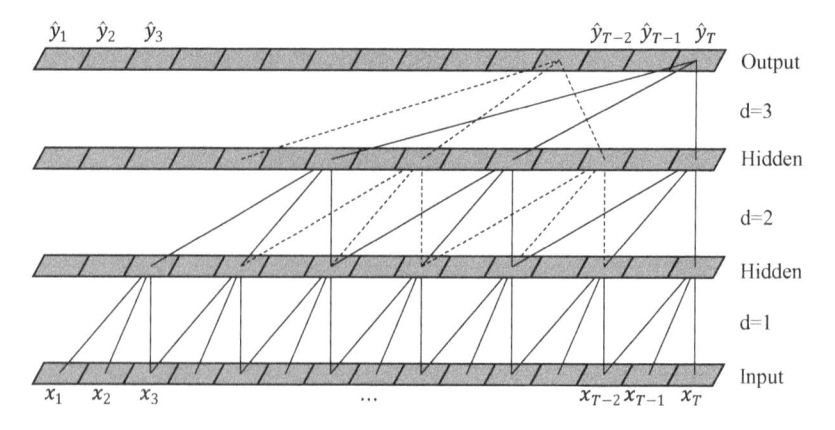

Fig. 2.16 Data flow in temporal convolutional network

2.2.5 Generative Adversarial Networks

Generative Adversarial Networks (GANs) [1] is a unique kind of generative architecture that draws inspiration from the zero-sum game in game theory. It consists of a pair of deep learning models, namely a generator and a discriminator, that undergo adversarial training, as illustrated in Fig. 2.17. The objective of the generator is to comprehend or acquire the distribution of real data and generate novel data samples. On the other hand, the discriminator's aim is to discern whether the data originates from real data distribution or if it is artificially generated by the generator. Thus, the discriminator functions as a binary classifier in this context. The two models engage in a competitive process, striving to enhance their respective performances until they converge to a Nash equilibrium. At this particular equilibrium point, the discriminator model is deceived approximately 50% of the time, signifying the successful generation of plausible instances by the generator model.

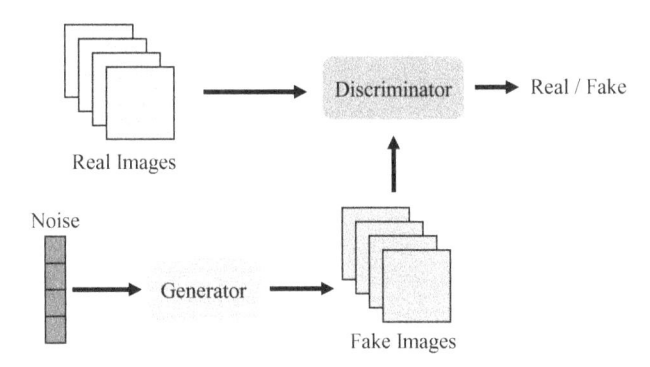

Fig. 2.17 The conventional process of image generation using Generative Adversarial Networks (GANs). Initially, a generator produces an image based on random noise, which is then evaluated by a discriminator. The discriminator assigns probabilities to both real and synthetic images, denoting authentic and fake images respectively, with 1 and 0. Subsequently, the discriminator output is utilized in a feedback loop with genuine images, and the generator output is looped with the discriminator to enhance the quality of the generated images

2.2.5.1 GAN Loss Function
The functioning of GANs can be described as follows: Consider two differentiable functions, G representing the generator and D representing the discriminator. The generator takes a random variable $\mathbf{z}$ as input and produces a data record $G(\mathbf{z})$,

learning the distribution p_g over the data $\mathbf{x}$ with a prior on input noise variables $p_\mathbf{z}(\mathbf{z})$. This generated record, along with real data records $\mathbf{x}$ sampled from the real data distribution $p_{\text{data}}(\mathbf{x})$, is fed to the discriminator for authenticity assessment. The discriminator assigns probabilities $D(\mathbf{x})$ to the records, with a prediction of 1 indicating a real data record and 0 indicating a fake record. Over time, the discriminator D is trained to maximize the probability of correctly classifying both training examples and samples generated by G. Simultaneously, G is trained to minimize $log(1 - D(G(\mathbf{z})))$. In essence, D and G engage in a two-player mini-max game with a value function $V(G, D)$ [1]:

$$\min_G \max_D V(D, G) = E_{\mathbf{x} \sim p_{data}(\mathbf{x})}[log D(\mathbf{x})] + \qquad (2.44)$$

$$E_\mathbf{z} \sim p_{\mathbf{z}(\mathbf{z})}[log(1 - D(G(\mathbf{z})))].$$

According to [1], the optimal discriminator $D_G^*(\mathbf{x})$ can be expressed as follows: As shown in [1], the optimal discriminator $D_G^*(\mathbf{x})$ is given by:

$$D_G^*(\mathbf{x}) = \frac{p_{data}(\mathbf{x})}{p_{data}(\mathbf{x}) + p_g(\mathbf{x})}. \qquad (2.45)$$

2.2.5.2 Boundary-seeking GAN

By rearranging Equation (2.45), we obtain the following expression:

$$p_{data}(\mathbf{x}) = p_g(\mathbf{x}) \frac{D_G^*(\mathbf{x})}{1 - D_G^*(\mathbf{x})}. \qquad (2.46)$$

Based on the above equation, it becomes evident that even if G is not optimal, the true data distribution can still be determined by scaling $p_g(\mathbf{x})$ [24]. Moreover, the optimal generator, where $p_{\text{data}}(\mathbf{x}) = p_g(\mathbf{x})$, can be attained by setting the discriminator ratio to 1. This implies that $D(\mathbf{x})$ must be equal to 0.5, which represents the decision boundary. In the case of a perfect G, $D(\mathbf{x})$ becomes unable to differentiate between real and fake data, making them equally probable. Since $D(\mathbf{x})$ has two outputs, each with a probability of 0.5, the objective function of G can be modified to compel the discriminator to output 0.5 for all generated data. Achieving this involves minimizing the distance between $D(\mathbf{x})$ and $1 - D(\mathbf{x})$ for all $\mathbf{x}$. As $D(\mathbf{x})$ represents a probability function, the minimum is reached when $D(\mathbf{x}) = 1 - D(\mathbf{x}) = 0.5$. Consequently, the loss function for generator G is defined as noted in [25]:

$$\min_{G} \mathbb{E}_{\mathbf{z} \sim p_{\mathbf{z}}(\mathbf{z})} \left[\frac{1}{2} (\log D(\mathbf{x}) - \log(1 - D(\mathbf{x})))^2 \right]. \qquad (2.47)$$

Our study utilizes the Boundary-seeking GAN due to its stable and efficient training characteristics.

2.3 Spiking Neural Networks

While ANNs were initially inspired by biological nervous systems, they are still unable to fully replicate the complex neurocomputational properties of biological neurons. To address this issue, the neuromorphic community has developed a third generation of ANNs known as SNNs. Unlike ANNs, SNNs closely emulate the functionality of the nervous system by incorporating both spatial and temporal aspects of input data into the computational model. The sparse and asynchronous communication in SNNs enables massively parallel data processing [24].

In addition, the energy-efficient implementation, fast inference, and event-driven information processing capabilities of SNNs make them an ideal candidate for executing deep neural networks and machine learning tasks that prioritize energy efficiency.

SNN research is still in its stages of development and which have been made possible by recent advancements in neuromorphic hardware platforms. However, there is still ambiguity on how to perform deep learning using these neurons. The existence of various types of spiking neurons without an agreed standard makes it a more challenging task and the designers of neuromorphic hardware must take this into account for researchers to evaluate model suitability [26]. With current advancements in SNNs, specialized accelerators and processors have been developed by both academia and industry to support these types of networks. Some research institutions and companies have even developed large-scale hardware solutions, such as SpiNNaker [27], IBM TrueNorth [28], and Intel Loihi [29] and Loihi 2 [30], for simulating spiking neural networks.

Although these processors are capable of simulating large-scale brain dynamics with reasonable accuracy, they are not optimized for ultra-low-power edge applications and consume a significant amount of energy. The processors are optimized for simulating SNNs, making them unsuitable for industrial applications that rely on deep learning. Researchers are currently searching for suitable event-based devices for industrial applications, and new hardware accelerators that rely on binary events computed in a synchronous manner are being developed as a midpoint between purely asynchronous SNNs and the synchronous processing of ANNs [26].

2.3.1 General Spiking Neural Network Architecture

SNNs are composed of interconnected neurons and synapses that communicate information via spikes. The synapses determine the propagation of information from a presynaptic (source) neuron to a postsynaptic (target) neuron [31]. While conventional Artificial Neural Networks (ANNs) use activation functions to encode information, SNNs transmit information through the activation of presynaptic neurons that alter the corresponding postsynaptic neurons. SNNs require multiple forward passes and are presented with input for a specified time period. When a presynaptic neuron activates, it generates a signal to its postsynaptic counterpart, which is proportional to the synapse's weight, changing the target neuron's membrane potential. If the membrane potential reaches a predefined threshold, the postsynaptic neuron fires a spike and resets its membrane voltage. SNNs provide various approaches to simulate the dynamics of neurons and synapses, and different network architectures and applications may require unique combinations of learning rules, neuron models, and membrane dynamics [31].

The SNN neuron models are presented at various levels of abstraction, ranging from the highly realistic and complex Hodgkin-Huxley (HH) model [32] to the simpler and computationally efficient leaky integrate-and-fire (LIF) model [33]. Among these models, the LIF model has gained substantial popularity due to its simple implementation, making it a computationally efficient choice while retaining important neurocomputational characteristics.

2.3.2 Leaky Integrate and Fire Model

The LIF is a modification of integrate-and-fire (IF) model by introducing a decay term that simulates the natural decay of neuron potential over time, thereby improving its biological plausibility. The LIF neuron is modeled as Resistor-Capacitor (RC) circuit as shown in Fig. 2.18. The mathematical formulation of the LIF model is given as [34]:

$$C_m \frac{dv}{dt} = g_L(V(t) - E_L) + I(t), \qquad (2.48)$$

2.3 Spiking Neural Networks

where C_m represents the membrane capacitance, g_L represents the conductance of the leakage channels, E_L represents the equilibrium potential of the leakage channels, V represents the voltage, t represents time, and I represents the total input current. The LIF model is capable of generating synchronous spikes to asynchronous inputs in certain firing patterns, mimicking the behavior of neurons in the human brain.

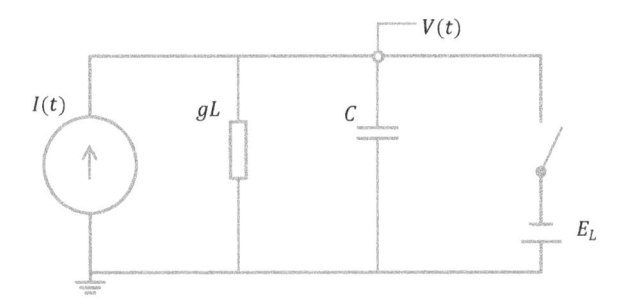

Fig. 2.18 Schematic representation of the LIF neuron model as a parallel RC circuit. The LIF neuron is modeled as an equivalent circuit consisting of a parallel combination of a resistor (g_L) and a capacitor (C_m), where the resistance represents the leak conductance and the capacitance represents the membrane capacitance. The circuit is driven by an input current (I) and generates output spikes when the membrane potential reaches the threshold voltage (V_{thres}). The parallel RC circuit model provides a simplified yet effective way to understand the basic functioning of the LIF neuron [31]

Due to the model's leaky behavior, V(t) continually decreases toward its resting value. Additionally, LIF neurons have a refractory period, which is a period after resetting during which the neuron cannot fire again, despite the input it receives. The behavior of a LIF node when exposed to a spike train input is illustrated in Fig. 2.19. Therefore, the activation function $h(t)$ is mathematically expressed as:

$$h(t) = \begin{cases} 0, & \text{if } V(t) < V_{thres}, \\ 1, & \text{if } V(t) \geq V_{thres}, \end{cases}$$

where V_{thres} is the firing threshold.

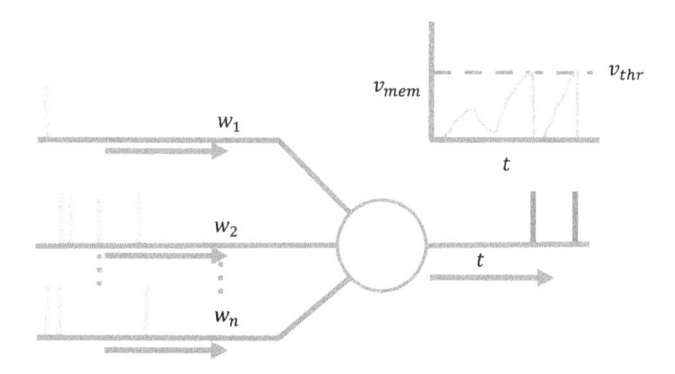

Fig. 2.19 Illustration of the behavior of a LIF-neuron when exposed to a spike train input, generates output spikes only after the refractory period has ended [31]

2.3.3 Encoding Schemes

Encoding schemes in SNNs refer to the techniques employed for representing information transmission within neurons through spike patterns. These schemes provide insights into the conversion of input data into spikes, enabling their processing within the SNN network.

There are two broad categories of encoding schemes rate-based encoding and temporal encoding [35].

2.3.3.1 Rate-based Encoding

Rate coding is one of the widely used encoding schemes in SNNs for representing information. Within this scheme, a neuron's firing rate serves as the basis for conveying information. The frequency of spikes generated by the neuron corresponds to the strength of the input signal. Notably, weaker input signals yield a lower firing frequency, while stronger input signals lead to a higher firing frequency [35].

Count Rate is the mean number of spikes released by a neuron within a particular time frame. If N_{spikes} is the spikes count in time window T then mathematically count rate is expressed as [35]:

$$F_r = N_{spikes}/T. \tag{2.49}$$

This scheme is also known as frequency coding.

Density Rate is computed by averaging the neural activity across multiple runs or simulations. To visualize the spiking activity, the results of the neural responses

2.3 Spiking Neural Networks

are plotted on a peri-stimulus-time histogram. Mathematically the spike density is given as follows [35]:

$$p(t) = \frac{1}{\Delta t} \frac{N_{spikes}(t; t + \Delta t)}{K}, \tag{2.50}$$

where the number of spikes is represented by N_{spikes}, the time interval over which the spikes are averaged is represented by $[t; t + \Delta t]$ for total K iterations. This approach is not biologically plausible for encoding information.

Population Rate is the mean activity level exhibited by a collection of neurons that possess similar properties [35]. Mathematically, the firing rate in this scheme is expressed as:

$$A(t) = \frac{1}{\Delta t} \frac{N_{spikes}(t; t + \Delta t)}{N}, \tag{2.51}$$

where N_{spikes} is the total number of spikes, within a time interval $[t; t + \Delta t]$ is summed across all neurons in the population. The sum is then divided by the duration, Δt, and the total number of neurons, N.

2.3.3.2 Temporal Encoding

In temporal encoding, the information is represented by the precise timing of individual spikes or action potentials, instead of relying on the rate or amplitude of a continuous signal [35]. Some of the widely used temporal encoding methods are illustrated in relation to a stimulus in Fig. 2.20.

Time-to-First-Spike (TTFS) is the simplest form of temporal coding, which represents information through the time difference, Δt, between the presentation of a stimulus and the first spike emitted by a neuron. The smaller the interval between stimulus onset and the initial spike, the larger the magnitude of the signal represented by the neuron. This approach is considered biologically plausible since it mimics the behavior of numerous neurons in the brain that exhibit such characteristics [35].

Rank-Order Coding (ROC) is a coding scheme that represents information by taking into account the firing order of a group of neurons relative to a global reference. In contrast to TTFS encoding, ROC does not take into account the precise timing of individual spikes. Instead, it acts as a discrete normalization filter, which leads to the loss of absolute amplitude information. Consequently, it becomes impractical to reconstruct the exact signal amplitude or a constant signal using ROC [35].

Inter-Spike Interval (ISI) also called latency coding, is a coding scheme where the information is sent through the timing of neuron signals. Instead of just looking at whether there's a signal or not, it pays attention to the time between signals. This helps the nervous system send information by understanding the specific time gaps

(a) Stimulus

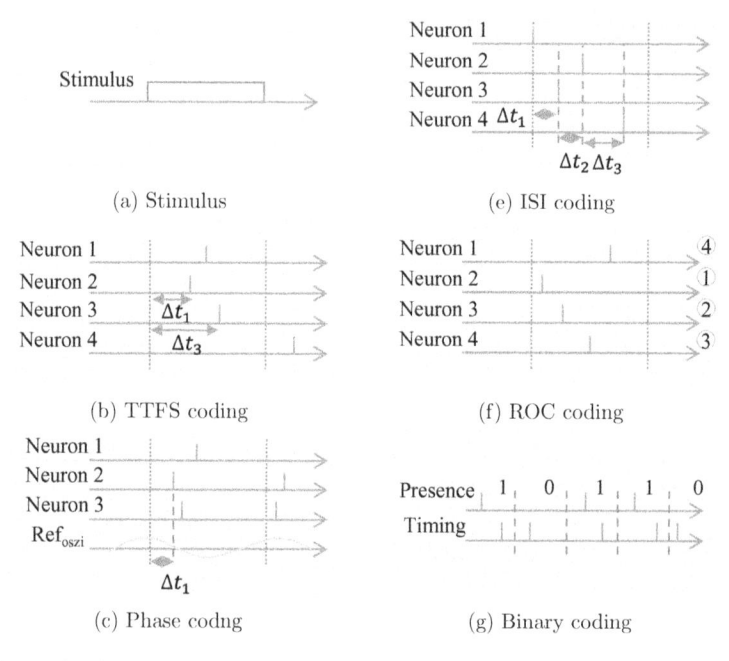

(b) TTFS coding

(c) Phase codng

(e) ISI coding

(f) ROC coding

(g) Binary coding

Fig. 2.20 The visualization showcases the utilization of temporal coding techniques, where a dashed line denotes the rising and falling edge of the stimulus. The parameter Δt represents the time difference between the reference point and the spike. In panel (f), the sequence of spikes is enumerated on the right-hand side for clarity [35]

between neuron signals. It's like finding a pattern in the timing of these signals to share information [35].

Phase Coding encodes information by measuring the relative time difference between spikes and a reference oscillation, rather than a single reference point. The phase pattern repeats periodically in the absence of any changes between cycles. Every neuron fires in correspondence to the reference signal, encoding data in a manner similar to Time-To-First-Spike (TTFS) coding [35]. Phase coding encodes information quantifying the relative temporal disparity between spikes and a reference oscillation, as opposed to a singular reference point. The encoded information manifests as a repeating pattern of phases in the absence of any modifications between cycles. Each neuron fires in synchronization with the reference signal, thereby encoding data in a manner similar to time-to-first-spike (TTFS) encoding [35].

Binary Coding involves the utilization of a stream of bits, where each spike represents either a "1" or "0". Two encoding schemes can be employed with respect to a fixed reference clock. The first scheme determines a logical "1" or "0" based on the presence or absence of a spike within a given interval. Alternatively, the second scheme encodes the bit by considering the timing of the spike within the interval. In this case, the clock cycle is partitioned into two sub-intervals, with a spike's presence in the first half indicating a "0", and in the second half indicating a "1" (or vice versa). By employing these schemes, a constant presence of spikes is ensured, irrespective of the encoded bit pattern [35].

2.4 Neural Engineering Object (Nengo) Simulator

Nengo [36], a neural engineering tool, is widely employed to simulate complex neural systems and has diverse applications in cognitive science, psychology, Artificial Intelligence, and neuroscience [37]. Its functionality is extended through the incorporation of NengoDL, a deep learning simulator seamlessly integrated with the TensorFlow library. This integration provides access to advanced capabilities, including convolution connections [38]. Nengo leverages a neural engineering framework (NEF) to design spiking neuron models specifically tailored for machine learning and deep learning applications. These models have showcased their effectiveness across various tasks, such as image classification [37], action selection [39], inductive reasoning [39], speech production [40], motor control [41], and planning with problem solving [42].

2.4.1 NengoDL

In this thesis, we used NengoDL [38], a tool that integrates with Nengo and facilitates the development of biologically plausible neural networks. NengoDL leverages the principles of the Neural Engineering Framework (NEF) to construct neuron models, enabling the representation, transformation, and dynamics required for building neural models with biological plausibility.

2.5 Legendre Memory Unit

In the domain of neural communication, synapses play a crucial role in transmitting and filtering spikes through intricate processes. These processes can be approximated using ordinary differential equations (ODEs) to capture the behavior of a cell over time [43]. The Legendre Memory Unit (LMU) is an approach that leverages this concept by effectively modeling continuous-time delays [44]. The LMU achieves this by determining appropriate weights for a set of Legendre basis functions, which are used to predict the output signal at a given time [43]. If the LMU network is adequately trained on the training data, the projected output signal should closely resemble the input signal. This bears similarity to the idea of using Fourier coefficients to reconstruct time-varying signals by combining weighted sine functions [45].

2.6 Federated Learning

Federated Learning (FL) is a decentralized approach for training a generic machine learning (ML) model by utilizing multiple local models trained on individual data from distributed clients [46], [47]. The FL algorithm involves two types of participants: "clients" and a "central coordinator" (referred to as the "server"). In this mechanism, each client possesses a local dataset consisting solely of its own data. The server shares a central model with all the clients. The clients enhance their respective models by training on their local data and subsequently transmit the updated training parameters to the server. The server then aggregates these parameter updates from multiple clients, generating an improved central model which is shared back with the clients. This iterative process continues as the clients gather more data. Consequently, FL enables learning from the client's data without the need for centralizing it.

Furthermore, FL offers several features that make it well-suited for distributed ML on mobile client devices. For instance, it can handle unbalanced and highly distributed datasets, process non-independent and non-identically distributed data, and exhibit robust performance even with limited communication capabilities.

2.7 Differential Privacy

Differential privacy (DP) is a method used to protect privacy in statistical databases by managing the trade-off between query accuracy and the potential disclosure of sensitive information. It ensures privacy preservation by introducing noise either to the output of statistical queries or the input data itself. The degree of differential privacy is quantified and assured by a privacy-loss budget represented as ε. A smaller value of ε corresponds to a higher level of privacy preservation, while a larger value allows for more accurate results but may compromise privacy to some extent. In this thesis, the introduction of DP is solely aimed at providing an understanding of the state-of-the-art pipeline for the forecasting use case. However, it does not delve into the privacy aspect of the use case. For an in-depth investigation of the privacy aspect in the mentioned use case, interested readers are directed to refer to [48].

2.7.1 Laplacian Noise Addition Mechanism

The Laplacian mechanism is a commonly used technique in DP [49] for introducing noise. Its purpose is to ensure DP-queries by adding random noise according to the Laplacian distribution. The magnitude of the noise is determined based on the sensitivity of the query function, denoted as S_f. In our specific approach, we adopt Laplacian differential privacy, where we directly apply noise to the aggregated data records.

In the traditional Laplace mechanism, random noise is generated from a Laplacian distribution with a mean of 0 and a variance of S_f/ε, aiming to achieve ε-differential privacy [50]. In our implementation, we incorporate individual noise for each data point within the aggregated data records. This is accomplished by selecting random noise samples from a Laplacian distribution with parameters $Lap(0, 1/\varepsilon)$. By employing this method, we ensure the desired level of privacy protection for the data.

Signal Processing Chain with Spiking Neural Networks for Radar-based Gesture Sensing

3.1 Introduction

Gesture sensing (GS) technologies enable the creation of touchless, non-invasive, and intuitive interfaces that allow users to interact with various devices without the need for physical contact or manual controls [51]. It has enabled a wide range of applications, including automotive, gaming, TVs, and smartphones. Traditionally, GS technology is dominated by vision-based systems where computer vision techniques are applied to rich datasets obtained with camera sensors [52–55]. Camera-based solutions, however, raise privacy concerns since they operate with photos and videos. Additionally, they require appropriate lighting conditions and weather conditions (in outdoor scenarios), which limits their application.

To overcome these limitations, non-vision based solutions have been developed. Non-visual systems typically involve attaching special sensors to people in the form of gloves or bands to record the movements of the human hand and recognize gestures by analyzing the data received from these sensors [56–60]. Although these systems have been shown to overcome the limitations of visual systems, wearing such sensors is very cumbersome for the user.

The limitations outlined above are overcome through the development of contactless non-vision-based systems. Radar-based technologies are particularly suitable for this, thanks to their subtleness under changing lighting conditions, invariant to the presence of hands, and simple signal processing capabilities [61–63]. Furthermore, its ability to maintain privacy, work inside enclosures, and detect fine motions makes it an ideal sensor for GS applications. Radar-based gesture sensing systems have two major research directions, one focusing on the development of efficient miniature hardware capable of producing high-fidelity target information. In radar-based gesture sensing systems, there are two main research directions: one focuses on developing efficient miniature hardware capable of producing high-fidelity target data [64–71], and the other focuses on the signal processing pipeline, where deep learning plays a dominant role [72–81].

With the current increasing demands for radars in IoT devices, radar-based techniques nowadays are focused on energy efficiency, a vital requirement in IoT [82–84]. In this regard, a tiny radar chip integrated into a mobile phone running a small CNN network has been demonstrated in [85]. Similarly for low-power wearable devices a 2D CNN coupled with a temporal convolutional neural network operating on range-frequency Doppler has been demonstrated in [86]. Robust detection and recognition of gestures have been shown in [87] by including additional features including range, Doppler, elevation, and azimuth as input to CNN followed by recurrent neural networks (RNNs). In [88], the power and computational efficiency are achieved by developing a tiny CNN for hand gesture recognition for embedded applications. To take advantage of the sensitivity of the radar to minute motion, [89–91] utilized micro-Doppler [92] signatures to capture micro-motion dynamics of the gestures to classify minute gestures (finger gestures). Although conventional deep learning techniques have dominated this field and have remained the best candidates for detection and recognition, the inference times energy efficiency remains a serious concern [93], especially for edge devices. The major cause of energy consumption is due to the Multiply-accumulate (MAC) operations between layers of the networks. Therefore, the research is mostly focused on reducing the MACs and is achieved by using smaller networks, weight quantization, and running techniques.

3.1.1 State-of-the-Art

The conventional processing pipeline for radar data is depicted in Fig. 3.1. This pipeline comprises several sequential steps. Firstly, it involves generating a 2D RDI by utilizing FFTs across the fast-time and slow-time dimensions. Following this, an algorithm is employed to detect and cluster targets and noise within individual cells. If multiple reflections originate from a single target, they are grouped together. After the detection and clustering stage, feature images such as Doppler spectrograms are extracted. These feature images are subsequently processed using a deep classifier,

3.1 Introduction

which may be implemented using either a convolutional neural network or long-short term memory, depending on the specific problem. The subsequent sections provide comprehensive details regarding each of these pipeline blocks.

2D Range-Doppler Images → Detection & Clustering Algorithms → Feature Images → Classifiers (CNN/LSTM) → Target Classification

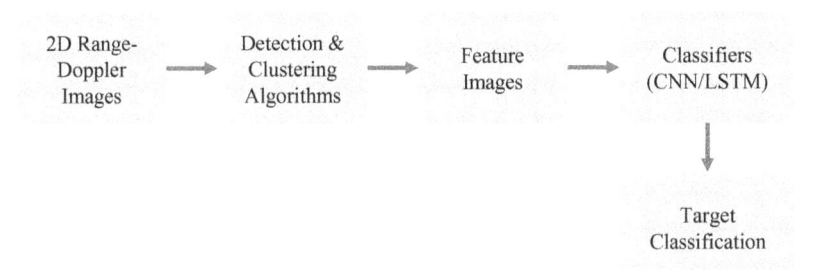

Fig. 3.1 Conventional radar processing pipeline for target detection and classification

3.1.1.1 2D Range Doppler Images

The signal received at frame N_K, denoted as $x_{IF}(t; N_K)$, from consecutive chirps is organized into a 2D matrix format as $x_{IF}(N_S, N_F; N_K)$. To generate the range-Doppler Image (RDI) for each channel, several steps are involved. Firstly, a window function is applied to the signal, followed by zero-padding to extend its length. Subsequently, a 1D FFT is performed along the fast time axis to obtain the range transformation. Further, another window function is applied, followed by zero-padding, and finally, a 1D FFT is performed along the slow time index. This process transforms the sampled signal $x_{\text{IF}}(N_S, N_F; N_K)$ into the range-Doppler domain using a 2D Discrete Fourier Transform (DFT). The transformation is mathematically represented as [20]:

$$X(p, q, N_K) = \sum_{N_S=1}^{Z_{N_c}} \left(\sum_{N_F=1}^{Z_{N_{TS}}} W_F(N_F) x_{IF}(N_S, N_F; N_K) \exp(-j2\pi p N_F / Z_{N_{TS}}) \right)$$
$$W_S(N_S) \exp(-j2\pi q N_S / Z_{N_c}), \tag{3.1}$$

where N_{TS} and $Z_{N_{TS}}$ represent the number of transmitted samples defined by DAC sampling points over the chirp duration and the zero-padding along the fast-time axis, respectively. Similarly, N_c and Z_{N_c} represent the number of chirps in a frame and the zero-padding along the slow-time axis, respectively. The terms $W_S(N_F)$ and $W_S(N_S)$ indicate the window functions applied along the fast-time and slow-time axes, respectively. The indices p and q correspond to the range and Doppler dimensions, respectively. The peaks in the range-Doppler domain are determined by the following equations [20]:

$$p_k = \left(\frac{2f_c}{c}v_k + \frac{2B}{cT}R_k\right),$$
$$q_k = \frac{2v_k f_c}{c}. \tag{3.2}$$

When employing a saw-tooth FMCW waveform with fast ramps, the condition $\frac{2f_c}{c}v_k \ll \frac{2B}{cT}R_k$ holds, indicating that the range peaks primarily appear at $\frac{2B}{cT}R_k$. The maximum achievable velocity is given by:

$$v_{\max} = \frac{c}{2f_c T_{\text{PRT}}},$$

and the minimum velocity is [20]:

$$\delta v = \frac{c}{2f_c Z_{N_c} T_{\text{PRT}}}.$$

Figure 3.2, illustrates the series of signal processing steps involved in generating RDIs from fast-time and slow-time ADC data. After applying the 2D FFT to convert the data into the range-Doppler domain, the subsequent step applies a moving average filter for background subtraction. Mathematically, this process can be expressed as follows:

$$X(p,q;N_K) = X(p,q;N_K) - X_B(p,q;N_K),$$
$$X_B(p,q;N_K+1) = \gamma X_B(p,q;N_K) + (1-\gamma)X(p,q;N_K), \tag{3.3}$$

where $X(p,q;N_K)$ represents the RDI at the N_K-th frame, while $X_B(p,q;N_K)$ corresponds to the background RDI at the N_K-th frame. The coefficient γ denotes the moving average coefficient utilized in the background subtraction process.

3.1 Introduction

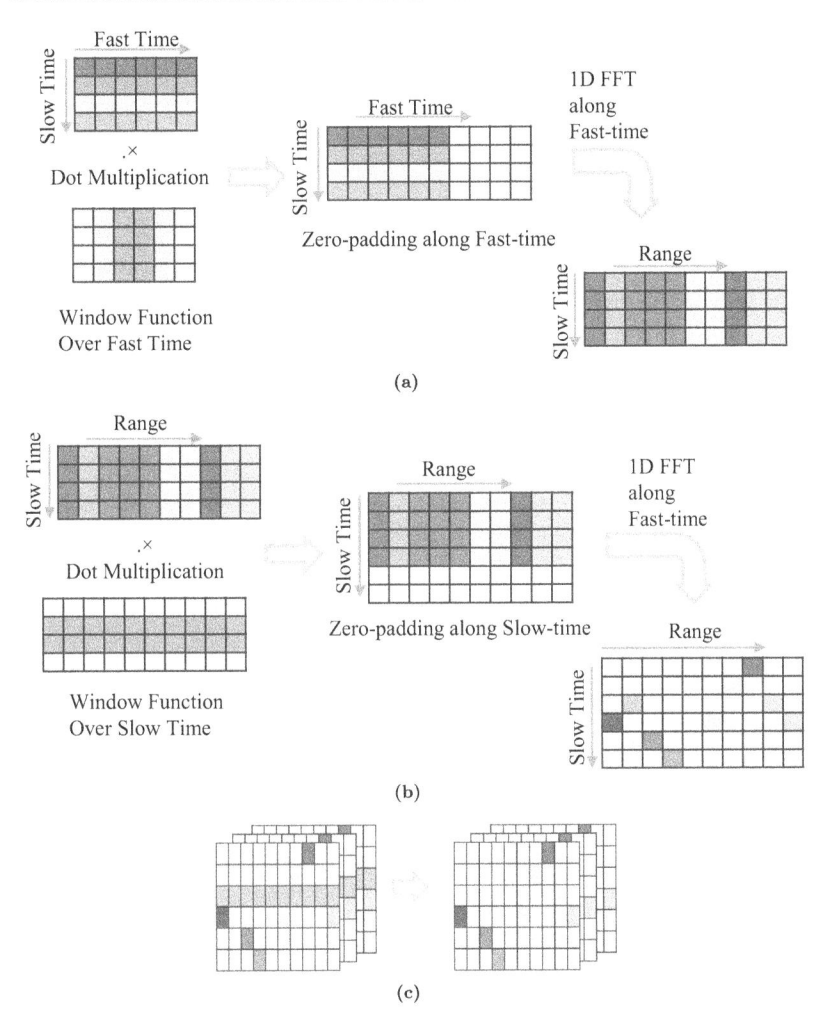

Fig. 3.2 (a) The steps involved in transforming raw data to range domain, (b) depicts the Doppler transformation steps, and (c) shows the background subtraction process

3.1.1.2 Detection Algorithm

After performing the transformation of the 2D range-Doppler or range cross-range data, a detection algorithm is applied to determine whether a cell under test (CUT) contains a target or not. Although it is feasible to apply this detection approach to the entire 3D data cube, it is often avoided when utilizing embedded processors due to the substantial computational complexity it entails. Figure 3.3 illustrates the application of the detection strategy on a 2D data matrix, which can be either an RDI, a range cross-range image, or an angle-Doppler image. The target detection problem can be formulated as follows [20]:

$$c_{\text{cut}} = \begin{cases} 1 & \text{if } c_{\text{cut}} > \mu \sigma_{cut}^2 \\ 0 & \text{if } c_{\text{cut}} \leq \mu \sigma_{cut}^2, \end{cases} \quad (3.4)$$

where c_{cut} represents the cell CUT. The threshold multiplier μ is typically utilized to ensure a fixed false alarm probability. Additionally, σ_{cut} denotes the power of the interference-noise bins, which is computed from the surrounding bins of the CUT.

Fig. 3.3 2D CFAR Detector

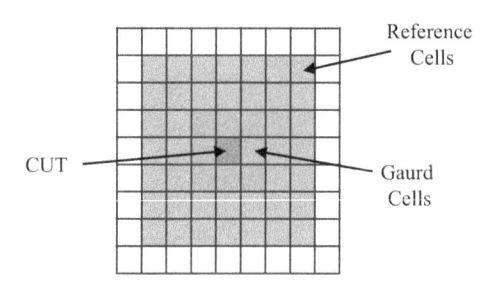

Radar detection algorithms are built upon the Neyman-Pearson criterion, which involves maintaining a constant probability of false alarm, denoted as P_{FA}. The primary objective of these algorithms is to maximize the probability of detection, denoted as P_D, under a given Signal-to-Noise Ratio (SNR) despite the presence of varying levels of interference or noise. These types of detectors are commonly known as constant false alarm rate (CFAR) detectors [20].

In environments such as indoor or automotive settings, the clutter levels tend to fluctuate due to the presence of interfering objects within the radar's field of view, as well as the existence of multiple targets. The detection algorithm operates under the assumption that the interference is independent and identically distributed across all resolution cells. Furthermore, the algorithm estimates the noise variance

3.1 Introduction

of the interference distribution to accurately differentiate between targets and interference [20].

One of the simplest forms of CFAR detection is known as the cell averaging CFAR algorithm. In this algorithm, the noise variance for the CUT is estimated by adding the squared values of the reference cells from both the lagging and leading positions. Mathematically, it can be expressed as [20]:

$$\sigma_{\text{cut}}^2 = \sum_{r=1}^{N/2} |c_{\text{lagging}}(r)|^2 + \sum_{r=1}^{N/2} |c_{\text{leading}}(r)|^2. \tag{3.5}$$

To establish the threshold multiplier, denoted as μ, to achieve a desired probability of false alarm. Specifically, in the case of the cell-averaging CFAR (CA-CFAR) technique, the threshold multiplier can be computed using the following formula:

$$\mu = N(P_{FA}^{-1/N} - 1), \tag{3.6}$$

where P_{fa} represents the probability of a false alarm, while N corresponds to the window size employed for estimating the noise power.

3.1.1.3 Clustering

In contrast to a point target, when dealing with doubly extended targets, the output of the detection algorithm does not yield a single detection in the RDI that corresponds to the target. Instead, the detections are distributed across both range and Doppler dimensions. Consequently, a clustering algorithm becomes necessary to group these detections from a single target into a coherent cluster based on their size. This clustering process proves beneficial in reducing the computational complexity of the target tracking algorithm. By clustering the detections, the tracking algorithm can focus on tracking a single target parameter, rather than dealing with non-clustered groups of target parameters [20].

A density-based spatial clustering of applications with noise (DBSCAN) algorithm, as proposed by [94], is a robust and effective unsupervised clustering technique. It aims to group target detections in a 2D space, accommodating both single and multiple targets. DBSCAN clusters detections that are closely packed together while removing outlier detections found in sparsely populated regions. The algorithm categorizes each point as a core point, edge point, or noise, based on specific criteria. To apply DBSCAN, two input parameters are required: the neighborhood radius, denoted as d, and the minimum number of neighbors, denoted as M. A point is considered a core point if it possesses at least $M - 1$ neighbors within a distance of d. An edge point has fewer than $M - 1$ neighbors but is adjacent to at least one core point. Any points with less than $M - 1$ neighbors and no core point neighbor are classified as noise, indicating that they do not belong to any cluster [94].

3.1.1.4 Radar Feature Images

Radar feature images work as two-dimensional representations of the target image, serving as inputs to a classifier tasked with categorizing the target based on motion/gesture type, object type, or pattern. Multiple types of two-dimensional radar features can be extracted to enable target classification, such as Doppler spectrogram and range angle images. Additionally, three-dimensional radar features, such as a sequence of range-Doppler images (RDIs) presented in video format, can also be utilized [20].

Doppler Spectrogram generations involves several steps. Initially, the target is detected along the range bins, and subsequently, an FFT is applied along the slow time axis for the detected range bin. The resulting Doppler spectra from consecutive frames are then stacked together to form a 2D image.

In the case of a single target, the Doppler spectrum can be obtained by aggregating the RDI data across the range axis. This Doppler spectrum comprises both the macro-Doppler component and the micro-Doppler components, which arise from the movements of hands etc. By stacking the Doppler spectra across consecutive frames, a Doppler spectrogram is created. This spectrogram provides valuable insights into the instantaneous Doppler spectral content and its variations over time [20].

The Doppler spectra of the slow-time data from the kth radar frame, specifically on the selected range bins, can be mathematically described as follows:

$$X(p,k)^n = \left| \sum_{q=1}^{N_{st}} w(q) s(q,k)^n \exp\left(-\frac{j2\pi q p}{N_{st}}\right) \right|,$$

$$X(p,k) = \sum_{n=1}^{N} X(p,k)^n, \qquad (3.7)$$

where $w(q)$ denotes the window function along the slow time, which is indexed by q. The variable $s(q,k)^n$ represents the slow-time data obtained across N_c chirps for the nth range bin in the kth frame. The parameter p refers to the number of FFT points along the slow time axis. The quantity $X(p,k)^n$ represents the Doppler spectrogram associated with the nth range bin, while N represents the total number of range bins. To generate the Doppler spectra for the given frame, the Doppler FFT

3.1 Introduction

is applied to the selected range bins, and the results are summed over the range bins [20].

Range Angle Images are acquired through several techniques. Among them, the straightforward method is applying a window function and zero-padding, followed by performing a Fast Fourier Transform (FFT) along the virtual channel.

To achieve this, the $N_t \times N_r$ de-ramped beat signal can be organized into a vector. By taking the Kronecker product of the transmit array's steering vector $a^{Tx}(\theta)$ and the receive array's steering vector $a^{Rx}(\theta)$, denoted as $A^{Tx}(\theta) \otimes A^{Rx}(\theta)$, the relative angle θ of the scatterer can be resolved. Here, the symbol $\otimes$ represents the Kronecker product. Subsequently, the process of beamforming the signals from the MIMO array can be interpreted as synthesizing the received signals using the transmit and receive steering vectors [20].

The azimuth imaging profile for a specific range bin l can be generated using the Capon spectrum obtained from the beamformer. The Capon beamformer is designed to minimize the variance or power of noise while maintaining an undistorted response towards a desired angle. This objective is achieved by solving a quadratic optimization problem defined as follows [20]:

$$\min_{w} w^H C w,$$
$$s.t. \quad w^H (A^{Tx}(\theta) \otimes A^{Rx}(\theta)) = 1, \quad (3.8)$$

where C represents the covariance matrix of noise. The solution to this optimization problem can be obtained in closed form as $w_{\text{capon}} = \frac{C^{-1} A(\theta)}{A^H(\theta) C^{-1} A(\theta)}$, where θ denotes the desired angle. By substituting the derived expression for w_{capon} into the objective function, the spatial spectrum can be obtained as follows:

$$P_l(\theta) = \frac{1}{\left(A^{Tx}(\theta) \otimes A^{Rx}(\theta)\right)^H C_l^{-1} \left(A^{Tx}(\theta) \otimes A^{Rx}(\theta)\right)} \quad (3.9)$$
$$\text{with } l = 0, ..., L,$$

where l ranges from 0 to L, representing different range bins. However, in practical scenarios, estimating the noise covariance at each range bin l is challenging. Therefore, an estimate $\hat{C}l$ is used, which includes both the signal and noise components. The sample matrix inversion technique is employed to estimate $\hat{C}l$, given by:

$$\hat{C}_l = \frac{1}{N} \sum_{k=1}^{K} s_l^{\text{IF}}(k) s_l^{\text{IF}}(k)^{\text{H}}, \qquad (3.10)$$

where K corresponds to the total number of snapshots utilized to estimate the covariance of the signal and noise. The symbol $s_l^{\text{IF}}(k)$ refers to the de-ramped intermediate frequency signal at range bin l, with k representing the frame index.

Video of Range-Doppler Images (RDIs) computed at frame time k can be represented by the following equation [20]:

$$\text{v}_{\text{RDI}}(p, l, k) = \left| \sum_{q=1}^{N_{st}} \sum_{n=1}^{N_{ft}} w(q, n) s(q, n, k) \exp\left(-j 2\pi \left(\frac{qp}{N_{st}} + \frac{nl}{N_{ft}}\right)\right) \right|, \qquad (3.11)$$

where $w(q, n)$ denotes the two-dimensional weighting function along the fast-time and slow-time dimensions, and $s(q, n, k)$ represents the ADC data on the kth frame. The indices n and m iterate over the fast-time and slow-time axes, respectively, while l and p iterate over the range and Doppler axes, respectively. N_{st} and N_{ft} refer to the FFT size along the slow-time and fast-time dimensions, respectively.

3.1.1.5 Deep Classifiers

In the last phase, a classification algorithm is employed to analyze the feature images or extracted features in order to determine the category or type of each target. Commonly utilized machine learning techniques, such as neural networks, are applied for this purpose. The output of the classifier can either provide the target's class directly or present a confidence score that indicates the probability of the target belonging to a specific category.

3.1.2 Limitations of State-of-the-Art

Despite the promising outcomes observed in detection and recognition with conventional deepNets approaches, the issue of energy efficiency persists. The primary cause of high energy consumption can be attributed to the multiply-accumulate (MAC) operations performed between layers. These operations involve complex computations, resulting in a notable drain on power resources.

Furthermore, the utilization of the FFTs in the radar signal processing chain contributes to energy inefficiency. The FFTs operation necessitates substantial computational capabilities and often requires additional hardware components for efficient execution. Consequently, this further amplifies the energy consumption, rendering such systems unsuitable for deployment on edge devices.

3.1.3 Proposed Solutions

In contrast to conventional deepNets, we adopted an alternative approach for radar-based gesture recognition systems, where we utilized SNNs in our proposed approaches to achieve energy efficiency. The recent advancement in SNNs [95] and the availability of resources to build specialized kinds of hardware to run these networks have made SNNs quite popular among the research community. Unlike deepNets, where information is typically transmitted through continuous numerical values, SNNs use the timing and frequency of spikes to communicate information, including the latencies and spikes rates. SNNs rely on sparse communication, meaning that information is only transmitted when a neuron's membrane potential reaches a certain threshold, resulting in the transmission of an action potential. In addition, the sparse communication nature of SNNs, which is characterized by 1-bit activity, leads to a substantial reduction in data volume that is transmitted between nodes. Moreover, since the nodes in SNNs perform a simple integration of incoming spikes, the multiplication-accumulation (MAC) arrays used in conventional neural networks can be replaced with adders, resulting in a significant decrease in computational requirements. Despite the energy efficiency of SNNs [96–99], the non-differentiable transfer function used in SNNs makes training them a difficult task.

Since the neuron's activation functions in SNNs do not facilitate conventional backpropagation, local unsupervised learning techniques such as Spike timing-dependent plasticity (STDP) and its variants are commonly utilized for training the SNNs [100]. But, these methods are only suitable for small networks with fewer parameters. While recent advancements in STDP have shown promising results [101], the concepts of deepNets have been incorporated for larger networks, utilizing differential approximations to train the spiking neurons in a backpropagation manner. The leaky integrate-and-fire (LIF) is the most commonly used spiking neuron model in SNNs due to its simplicity, ease of implementation [102], low computational requirements, and neuro-computational characteristics.

The main contribution of this chapter are as follows:

- In Sect. 3.4 we propose a novel gesture classification system with the following contributions:

- We propose a novel SNN architecture that utilizes range-Doppler information to classify gestures.
- To the extent of our knowledge, this is the first study that investigates the feasibility of SNN for radar gesture sensing.
- In Sect. 3.5 we propose a novel gesture classification system with the following contributions:
  - We propose a novel classification architecture that operates without a convolutional layer, thus fulfilling the hardware requirements. This was made possible by refining the SNN architecture.
- In Sect. 3.6 we propose a novel gesture classification system with the following contributions:
  - We have added more complex gestures, up to 8 different gestures, for a comprehensive evaluation of the proposed solution.
  - A thorough evaluation of additional spiking neuron models and their behavior is performed on the augmented dataset.
  - Enhancements have been made to the preprocessing pipeline in order to produce more effective and representative features that can be learned by the predictive network.
  - We propose a modified and improved SNN architecture that can accurately detect and identify all 8 gestures with a high degree of precision.
  - In contrast to range-Doppler images, more comprehensive hand profiles, namely range-time, velocity-time and angle-time images are utilized as features to improve the robustness of the system.
- In Sect. 3.7 we propose a novel gesture classification system with the following contributions:
  - We propose an end-to-end pipeline for a radar-based gesture sensing system that works solely on raw data using SNN.
  - In contrast to prior methods in Sects. 3.4, 3.5 and 3.6 that operate on Doppler images, our proposed approach works directly with the raw ADC data, eliminating the need for slow-time and fast-time FFTs and thereby reducing overhead.
  - We propose a novel SNN architecture in which the slow-time FFT signal preprocessing is emulated within the SNN.
  - Compared to previous approaches Sects. 3.4 and 3.5, the overall latency from start to end has been significantly reduced by a factor > 2.
- In Sect. 3.8 we propose a novel gesture classification system with the following contributions:

3.2 Hardware

- A gesture sensing system based on an end-to-end SNN is presented, wherein the SNN directly processes the raw data and accomplishes the task of classification.
- In contrast to the methods proposed in Sects. 3.4, 3.5 and 3.6, which operate on Doppler images, the proposed approach exclusively utilizes raw ADC data. This eliminates the need for pre-processing steps like slow-time and fast-time FFTs, resulting in a reduction of the computational overhead required for execution.
- The proposed approach advances the proposed approach in Sect. 3.7 that only classifies 4 gesture, the proposed approach not only mimics the fast-time FFT but also slow-time FFT in SNN enabling the system to classify up to 8 gestures.
- The proposed approach advances the method presented in Sect. 3.7, which primarily focused on classifying only four gestures. In contrast, the proposed approach goes beyond mere mimicry of the slow-time FFT and also incorporates the fast-time FFT within the SNN framework. This enhancement significantly enhances the system's classification capabilities, enabling it to effectively classify up to eight different gestures.
- A novel SNN architecture is introduced, which mimics the signal pre-processing steps, namely the slow-time FFT and fast-time FFT, within the SNN framework.
- In Sect. 3.9 we develop an FPGA demonstrator for gesture sensing. Our Xilinx FPGA implementation is able to achieve a comparable level of performance in terms of accuracy 98.5%.

The rest of the chapter is organized as follows: Section 3.2 presents the radar hardware used in this chapter, and Sect. 3.3 presents the dataset collection and experimental setup. The Sects. 3.4, 3.5, 3.6, 3.7 and 3.8 presents our proposed systems.

3.2 Hardware

In this chapter, the FMCW radar chipset *BGT60TR13C* by Infineon Technologies, as depicted in Fig. 3.4(a), was utilized. The simplified block diagram of the chipset can be seen in Fig. 3.4(b). The chipset comprises of a transmitter, receiver, mixer, and Analog-to-Digital Converter (ADC), alongside an external phase-locked loop responsible for regulating the linear frequency sweep. The frequency divider output pin, operating with an 80 MHz reference oscillator, governs the control loop. To synchronize the Finite State Machine (FSM) [103], a reference clock

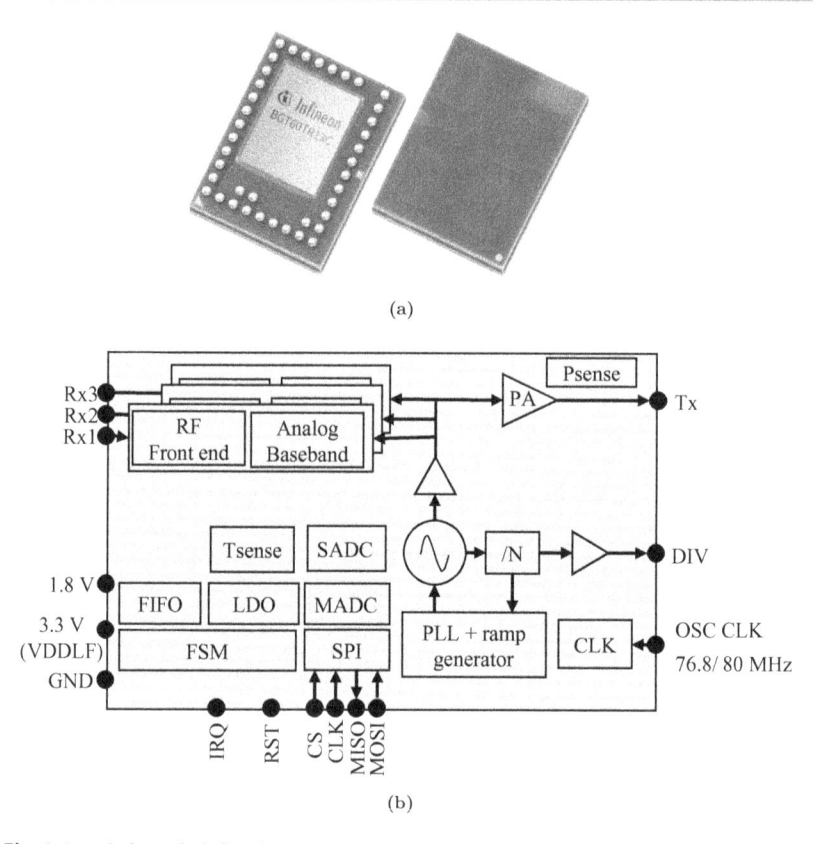

Fig. 3.4 a) Infineon's BGT60TR13C FMCW radar chipset utilized in this study, b) and its simplified block diagram

operating at 80 MHz is employed. To achieve linear frequency sweeps spanning from 57 GHz to 63 GHz, the voltage-controlled oscillator (VCO) is activated by adjusting the tuning voltage V_{tuned} within the range of 1 to 4.5 V. The chipset consists of one transmit (Tx) path antenna and three receive (Rx) paths. It is equipped with both Serial Peripheral Interface (SPI) and Queued Serial Peripheral Interface (QSPI) for memory readout. The maximum data transfer rate supported is up to 200 Mb/s (4 × 50 Mb/s). To facilitate data streaming, the FSM triggers an interrupt (IRQ) flag as soon as the host-defined threshold in the memory is reached. The active high-pass filter is driven by the mixer, followed by a VGA, AAF, and an ADC driver. The multi-channel ADC (MADC) employed utilizes a 4 Mb/s 12 b

3.2 Hardware

Table 3.1 Radar parameters employed in our experimental setup

Parameter	Symbol	Value
Number of ADC samples	NTS	64
Number of Chirps	PN	32
Chirp Time	T_c	32 µs
Number of Transmit Antennas	N_{TX}	1
Number of Receive Antennas	N_{RX}	3
Total Bandwidth	B	5 GHz
Frame Time	T_f	75.476 ms
Azimuth Antenna Field of View	θ_{FOV}	70°

Successive Approximation Register (SAR) topology. An SRAM with a capacity of 196 kbit is employed for storing the raw data. Additionally, a sensor ADC (SADC) is utilized to obtain temperature and transmit power readout information [103]. By performing time domain multiplication, the transmitted signal generated by the TX is combined with the received signals from the RX. These resultant signals are then forwarded for subsequent processing. The chipset is capable of transmitting signals with a bandwidth of up to 6 GHz, thereby providing resolution Δr and Doppler resolution Δv as follows:

$$\Delta r = \frac{c}{2B} = 2.5 \, \text{cm}, \tag{3.12}$$

$$\Delta v = \frac{c}{2f_c} \cdot \frac{1}{N_c T_c} = 122 \, \text{cm/s}, \tag{3.13}$$

where c represents the speed of light, B denotes the bandwidth, f_c corresponds to the center frequency set at 60 GHz, N_C indicates the number of chirps set to 64, and the chirp duration T_c is established as $32 \mu s$.

3.2.1 System Parameters

In Fig. 3.5, the experimental setup is illustrated where a hand gesture is performed in the vicinity of the radar. The system parameters and derived parameters as outlined in Table 3.1 are used to configure the setup.

The raw ADC data is acquired from radar via a USB interface and stored in PC for further subsequent processing. Approximately, the distance between the hand and the radar is kept in the range of 0.2–1 m. Hand detection is done automatically

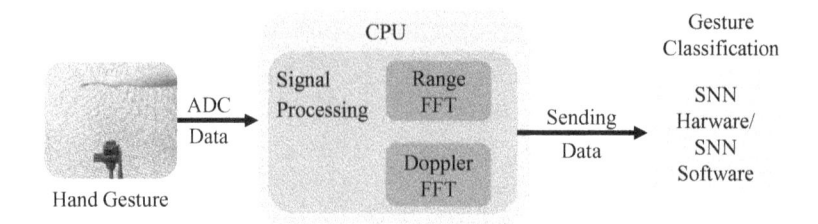

Fig. 3.5 The proposed system flow involves feeding raw radar data into a signal processing block, where fundamental tasks like range FFT and Doppler FFT are performed. The resulting data is then passed to a gesture classification block that employs an SNN to classify the gestures

by the system as soon as it enters the radar's field of view (FoV). At this point, the system designates it as the starting point for the gesture and initiates the frame count. For each gesture, the data is acquired for 32 frames where 32 are the name of frames taken by the slowest gesture. Every gesture is recorded and labeled with significant variation between different classes. If a gesture comprises less than 32 frames, it is appended with zeros. Additionally, to improve the latency of the gesture detection system, not only can the beginning of a gesture be detected, but also the end of the gesture can be identified using the same MSE method proposed. After the gesture's end is detected, for example, at 12 frames, zeros are appended to the spectrograms before they are input into the SNN for classification. This would improve the latency of the real-time solution, but it would not affect the detection and classification performance.

3.2.2 Signal Model

The functioning principle of FMCW radar involves the transmission of chirps, which are waves with a frequency that increases linearly. When reflected by a target, these chirps are collected at the receiver antenna. At the receiver, a mixer combines the transmitted and received signals. This mixed signal is low-pass filtered to get a constant frequency signal referred to as the Intermediate Frequency signal. Mathematically, the frequency of the waveform or chirp for bandwidth B_w and duration T can be expressed as:

$$v(t) = v_c + \frac{B_w}{T}t, \quad (3.14)$$

where v_c is the start frequency of the ramp. The beat signal is generated by mixing the reflected signal with a replica of the transmitted signal. The resulting down-converted Intermediate Frequency (IF) signal is given by:

$$I_F(t) = \sum_{m=1}^{M} \exp\left(2\pi\left(\frac{2v_c R_m}{c} + \left(\frac{2v_c v_m}{c} + \frac{2B R_m}{cT}\right)t\right)\right), \quad (3.15)$$

with the assumption of $\tau_m/T << 1$. Where $\tau_m = \frac{2R_m + v_m t}{c}$ represents the duration it takes for the transmitted signal to reflect from the m-th target positioned at a distance R_m from the radar, accounting for the radial velocity v_m. Here, the speed of light is denoted as c. The resulting down-converted Intermediate Frequency (IF) signal, denoted as $I_F(t)$, is sampled and prepared for further processing.

3.3 Dataset and Experimental Setup

The dataset consists of 4800 hand gesture swipes collected from 5 individuals and includes 8 distinct gestures as shown in Fig. 3.6. These gestures are:

1. Down-up swipe (moving the hand from down to up)
2. Up-down swipe (moving the hand from up to down)
3. Left-right swipe (swiping the hand from left to right)
4. Rubbing (rubbing two fingers together)
5. Right-left swipe(swiping the hand from right to left)
6. Diagonal-southwest-northeast swipe (swiping the hand from the left bottom corner to the right top corner)
7. Diagonal-southeast-northwest swipe (swiping the hand from the right bottom corner to the left top corner)
8. Clapping (clapping two hands)

The dataset was balanced with each gesture having 600 samples. The users were given minimal supervision during dataset collection. Furthermore, the dataset was gathered under diverse environmental conditions, including various locations and surroundings. To introduce additional diversity to the dataset, each gesture was demonstrated to two out of the five subjects, allowing them to perform the gestures based on their personal interpretation. As a result, their gestures exhibited less rationality compared to the three subjects who received minimal supervision. For training purposes, we randomly selected 80% of the dataset, while the remaining 20% was allocated for testing. This process was repeated across multiple trials.

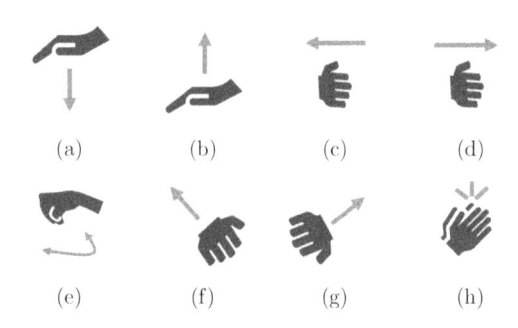

Fig. 3.6 Examples of the various types of gestures in the dateset, namely: a) Up-down gesture, b) Down-up gesture, c) Left-right gesture, d) Right-left gesture, e) Rubbing gesture, f) Diagonal southeast to northwest gesture, g) Diagonal southwest to northeast gesture, h) Clapping gesture

3.4 Gesture Sensing Using Range-Doppler Images

In this section, we introduce an innovative gesture recognition system based on SNNs, incorporating a 60-GHz FMCW chipset. The proposed approach leverages the range and Doppler information obtained from gestures and achieves comparable performance to deepNets in classifying four distinct gestures. The content covered in this section corresponds to our publication P11.

3.4.1 Signal Processing Chain

The raw data obtained from the radar is first fed to the signal processing block, which performs the fundamental signal processing tasks such as range FFT and Doppler FFT followed by the gesture classification block that performs the gesture classification using SNN as shown in Fig. 3.5. To incorporate the concept of time into SNN, we generate a range-Doppler image (RDI) for each frame and subsequently extract the vector associated with the highest value in the RDI (referred to as the RD vector). This vector is then inputted into the target classification block as shown in Fig. 3.7. To obtain the RDI for a single frame, the raw radar signal undergoes several steps. Initially, the signal is adjusted to have a zero mean, followed by the application of a Hann window multiplication in the fast-time direction. Next, the target's range is determined by performing a first-order FFT along the fast-time axis, utilizing an FFT size of 128. The calculation of the target's velocity relative to

3.4 Gesture Sensing Using Range-Doppler Images

the radar is achieved by conducting an FFT with a size of 64 along the slow-time direction, using only the positive values of the FFT. Collectively, these steps yield a single RD of dimensions 64×32.

3.4.1.1 Target Detection

Target detection and selection are performed using a simple thresholding technique. The threshold value α for frame k is determined by computing the mean value of the range FFT spectra and applying a scaling factor. Thus, the threshold value α for frame k can be computed as:

$$\alpha_k = \beta \sum_{n=1}^{N_s} R_{ci}^n(k), \qquad (3.16)$$

where β is a scaling factor set to 3 in this case, and n represents the index along the range bins.

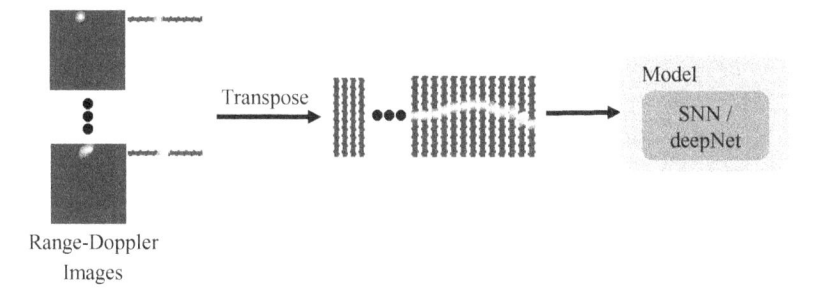

Fig. 3.7 Signal processing steps showing the transformation of range-Doppler image (RDI) to range-Doppler vector (RD) and then fed to the model. A RDI for each frame is created and then the vector associated with the highest value (RD vector) is extracted. The RD vector is then inputted into the target classification block

3.4.1.2 Gesture Classification

The RD vectors are fed into the gesture classification block, where SNNs are employed for classification purposes. The SNN block can be implemented using either a simulator or dedicated hardware, such as SpiNNaker [27], which is specifically designed for running SNNs. In our experiments, we utilized a simulator named Nengo [38].

3.4.2 Architecture & Learning

The primary aim of the proposed model is to maintain biological plausibility while minimizing computational complexity. To achieve this goal, the model utilizes the LIF neuron, which is biologically plausible and requires less computation.

3.4.2.1 Architecture

The architecture of the SNN is illustrated in Fig. 3.8. NengoDL [38] was utilized for constructing the network, which provides a differential approximation of the firing rate of LIF neurons. The input layer of the network comprises 64 × 1 nodes, followed by a convolutional layer with a filter size of 3, a total of 32 filters, and a stride of 2. After the convolutional layer, the output is converted into spikes using LIF as a non-linear function. This is then followed by a dense layer with 4 neurons, with LIF being applied to each of them. The last output layer of the network is also a dense layer of size 4. In training, the network employs SoftLIF [104] activation (an approximation to LIF) and utilizes a multi-class cross-entropy function as the objective function.

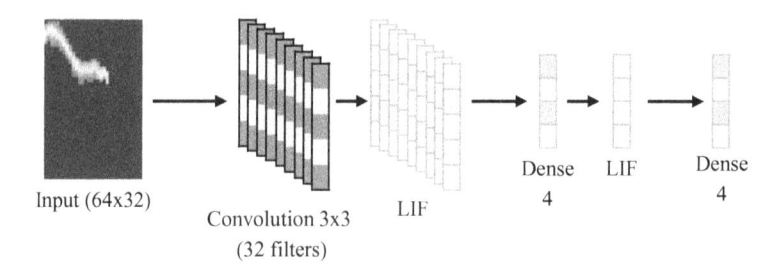

Fig. 3.8 Proposed spiking neural network architecture, consisting of an input layer with 64 × 1 nodes, a convolutional layer with a filter size of 3, 32 filters, and a stride of 2, a dense layer with 4 neurons, and a dense output layer of size 4. After the convolutional layer, the output is converted into spikes using LIF as a non-linear function. The network is trained using SoftLIF activation and a multi-class cross-entropy objective function

3.4.2.2 Model Testing

After training the network, LIF neurons are employed to reconstruct the model, effectively transforming it into an SNN. The training parameters of the model, including weights and biases, obtained from training with a differential approximation of LIF, are utilized to establish connections among spiking LIF neurons, forming a testing model. During the testing phase, the test inputs or samples are

3.4 Gesture Sensing Using Range-Doppler Images 65

presented to the network 25 times or steps, allowing for accurate measurement of the spiking neuron's output over time.

3.4.3 Results & Discussion

3.4.3.1 Dataset

Four gestures were chosen from the dataset described in Sect. 3.3. These four gestures include:

1. Top-down gesture (moving the hand in a downward direction)
2. Down-up gesture (moving the hand in an upward direction)
3. Right-left gesture (moving the hand towards the left)
4. Rubbing two fingers together

Figure 3.9 illustrates an example of one of these gestures along with its corresponding RD vector's map.

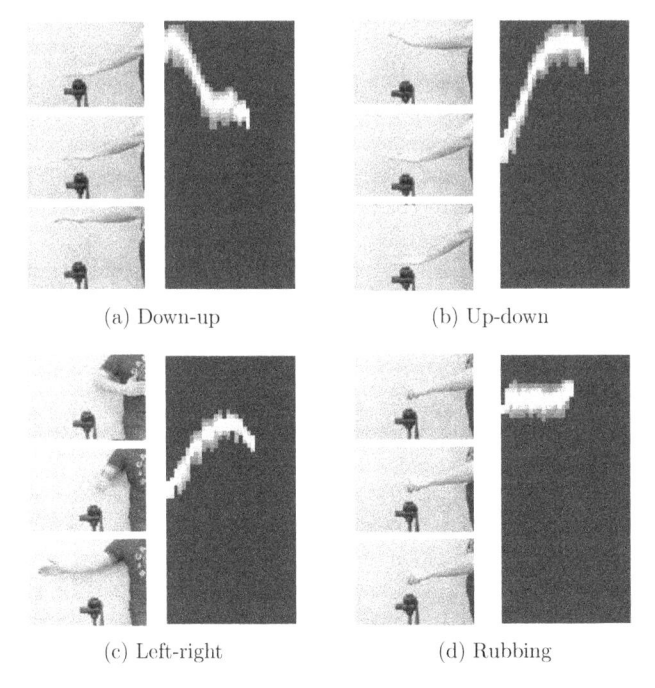

(a) Down-up   (b) Up-down

(c) Left-right   (d) Rubbing

Fig. 3.9 An example of a gesture along with its RD map of four gestures, selected from the dataset in Sect. 3.3. Which include moving the hand in a top-down or down-up direction, moving the hand towards the left, and rubbing two fingers

Table 3.2 Comparative analysis of classification accuracy: proposed SNN model versus other existing models

Approaches	Accuracy
Proposed model_1	**98.5%**
LSTM	96.9%
CNN-LSTM	97.18%

3.4.3.2 Classification Results

To evaluate the classification performance of the proposed system, accuracy was employed as an evaluation metric. The system consistently achieves a comparable average accuracy level to its deepNet counterparts across multiple random trials (Table 3.2).

3.4.3.3 Discussion

This section demonstrates the significance of the shift in paradigm from second-generation to third-generation neural networks. The new paradigm emphasizes not only static values but also the timing of those values [105]. In this section, we propose a gesture recognition system that is both resource-efficient and based on SNN, a third-generation neural network that employs fewer neurons than ANNs. Additionally, SNNs operate at a faster speed, consume less energy, and are easy to implement on hardware, making them a more cost-effective solution.

Earlier studies on gesture sensing using radar technology employed classical ANNs for classification but did not utilize temporal information. Similar to conventional CNNs, the effectiveness of SNNs is dependent on the choice of hyperparameters. In order to determine the optimal hyperparameters for training and testing the SNN, a grid search was conducted. The grid search led to the selection of the hyperparameter shown in Table 3.3.

The proposed system demonstrates accurate classification of four gestures with a level of 98.5% accuracy. This level of accuracy is comparable to conventional deepNets such as LSTM and CNN-LSTM, as demonstrated in Table 3.2. The ability of the proposed system to achieve this level of performance is credited to the utilization of spatiotemporal information encoding by SNNs, which leverages network dynamics for learning. In Fig. 3.10, the firing patterns for the four different gesture samples can be observed. It is evident that the SNN begins firing for a sample from the correct class after a few time steps. As the proposed system is based on SNN, it is anticipated to be more energy-efficient and faster in comparison to its deepNets equivalent.

3.4 Gesture Sensing Using Range-Doppler Images

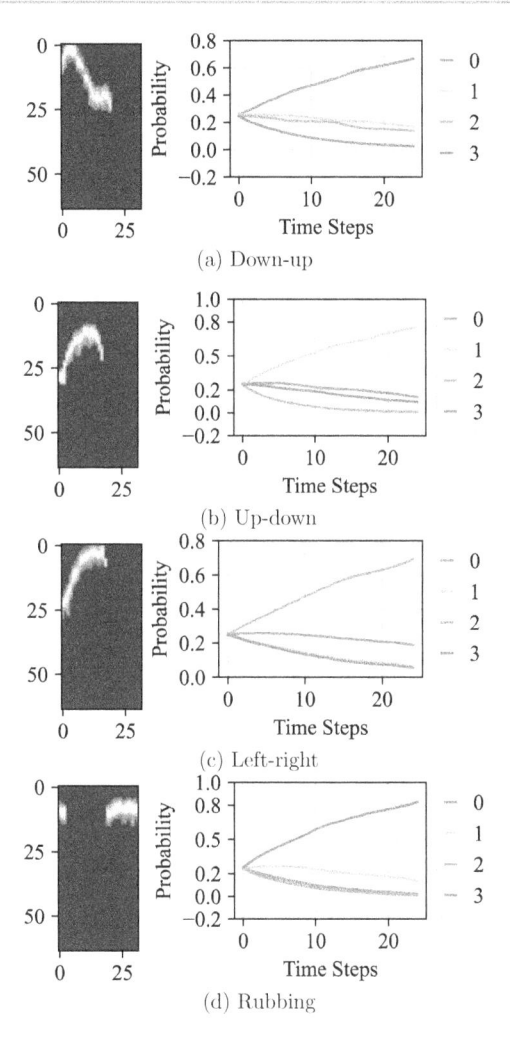

(a) Down-up

(b) Up-down

(c) Left-right

(d) Rubbing

Fig. 3.10 Firing patterns for the four different gesture samples are shown where the SNN starts firing for a sample from the correct class after a few time steps

Table 3.3 The hyperparameters used for configuring the proposed SNN

Parameters	Value
Learning rate (η)	0.001
Output synaptic delay	10 (ms)
Numer of layers	2
Number of filters in each layer	32
Neuron model type	LIF
Neuron amplitude	0.01
Batch size	100
Number of epochs	10
Number of time steps during testing (repetition of input)	25

3.4.4 Conclusion

In this section, we present a resource-efficient gesture recognition using FMCW radar. The proposed system leverages SNNs, enabling effective deployment on edge devices with minimal form factor and power usage. A notable challenge in utilizing SNNs lies in determining suitable learning rules to capture spatiotemporal spike patterns. We demonstrate that SNNs can achieve performance on par with conventional deepNet methods for 4 distinct gestures.

3.5 Optimized Gesture Sensing Using Range-Doppler Images

In this section we propose a novel SNN classification architecture that operates without a convolutional layer, thus fulfilling the hardware requirements. This was made possible by refining the SNN architecture. The content covered in this section corresponds to our publication P10.

3.5.1 Signal Processing Chain

We used a similar signal processing chain as elaborated in Sect. 3.4.1.

3.5 Optimized Gesture Sensing Using Range-Doppler Images 69

3.5.2 Architecture & Learning

The primary objective of the proposed model is to minimize computational complexity. To achieve this, we have opted to use the LIF approach, which not only requires less computation but is also biologically plausible. The architecture of the SNN is illustrated in Fig. 3.11. For constructing the network, we utilized NengoDL [38], which provides a differential approximation of the LIF neuron firing rate. The training process follows a conventional end-to-end learning approach using backpropagation.

The input layer of the network comprises a 64 × 1 node, which is subsequently connected to a dense layer consisting of 32 neurons. The dense layer is further coupled with LIF as the non-linear activation function, which maps the output to spikes. A dense layer with 4 neurons, followed by LIF, is then added to the network. This layer is succeeded by a final dense layer that functions as the output layer. During training, SoftLIF activation (an approximation to LIF) is used, with a multi-class cross-entropy function serving as the objective function.

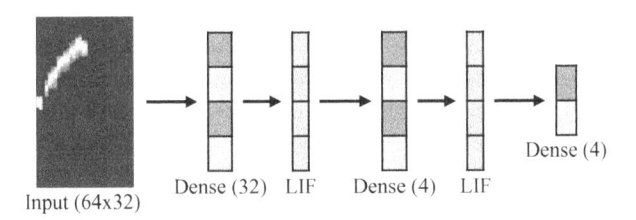

Fig. 3.11 The proposed SNN architecture includes an input layer of 64 × 1 nodes, two dense layers of 32 and 4 neurons coupled with LIF as the activation function, and an output layer. During training, SoftLIF activation and backpropagation are used. For testing, SoftLIF is replaced with LIF to create the SNN and the training parameters are used to connect spiking LIF neurons

3.5.2.1 Loss Function

During the classification, the probabilities of each class are determined using a softmax classifier. This classifier utilizes cross-entropy as the loss function, which is defined as:

$$L = -\frac{1}{K} \sum_{c=1}^{C} \sum_{k=1}^{K} y_k^c \log (h_\theta (x_k, c)), \quad (3.17)$$

where K represents the number of training examples, C is the total number of classes, y_k^c represents the target label for training example k for class c, x represents the input for the training example, and h represents the model with weights θ.

3.5.2.2 Model Testing

During the testing phase, the model is reconstructed with SoftLIF replaced by LIF to create the SNN. The training parameters (weights and biases) from the previously trained SoftLIF model are then used to connect spiking LIF neurons for testing. To obtain an accurate measurement of the spiking neuron output over time, the test inputs are modified to be presented multiple times or steps to the network.

3.5.3 Results & Discussion

3.5.3.1 Dataset
We used a similar dataset as described in Sect. 3.4.3.1.

3.5.3.2 Classification Results
The performance of the proposed system is evaluated using accuracy as a measure of classification performance. Table 3.4 demonstrates that the proposed system achieves a comparable level of accuracy to that of conventional deepNets and to the proposed model (Proposed model_1) in Sect. 3.4.

Table 3.4 Comparative analysis of classification accuracy: proposed SNN model versus other existing models

Approaches	Accuracy
Proposed model_2	97.5%
Proposed model_1	98.5%
LSTM	96.9%
CNN-LSTM	97.18%

3.5.3.3 Discussion
We demonstrate the significance of 3rd generation neural networks, which go beyond focusing solely on static values, as was the case with 2nd generation networks. Instead, 3rd generation networks take into account the occurrence times of these static values [105] into account as well. Our proposed gesture recognition system

3.5 Optimized Gesture Sensing Using Range-Doppler Images

is designed to be resource-efficient in terms of power consumption, using fewer neurons than ANNs by leveraging the 3rd generation neural network architecture. Furthermore, the use of SNNs provides advantages in terms of speed, scalability, and hardware implementation, resulting in a more cost-effective solution.

Similar to conventional CNNs, the performance of SNNs heavily relies on the appropriate selection of hyperparameters. In order to achieve optimal performance during both training and testing, a grid search was conducted to explore various hyperparameter combinations. This iterative process involved testing different combinations, allowing the network to potentially overfit or underfit, until the optimal set of hyperparameters, as depicted in Table 3.5, was determined. As demonstrated in the table, the proposed system achieves a commendable accuracy level of 97.5% in classifying four gestures. This performance is comparable to conventional deepNets such as LSTM and CNN-LSTM, as well as our Proposed model_1. The impressive performance of the proposed system can be attributed to its utilization of spatiotemporal information encoding through SNN. This approach effectively harnesses network dynamics for enhanced learning capabilities. Figure 3.13 shows the confusion matrix obtained from testing the proposed SNN using the dataset. The analysis reveals that the down-up gesture is frequently misclassified as finger rub, with a significant number of confusions occurring between down-up and right-left gestures. This confusion arises due to the similarity in the RD vector's map of down-up and right-left gestures, contributing to misclassifications. In Fig. 3.12 columns (b) and (c) illustrate the RD vector's map of four distinct gesture samples along with their firing patterns generated by the proposed SNN model. As shown, after a few time steps, the SNN begins to fire for a sample from the correct class. This is due to pre-

Table 3.5 The hyperparameters used for the proposed SNN

Parameters	Value
Learning rate (η)	0.001
Output synaptic delay	10 ms
Type of neuron model	LIF
Neuron amplitude	0.01
Batch size	100
No. of epochs	10
Time steps during the testing	25
No. of layers	2
No. of Neurons	32,4

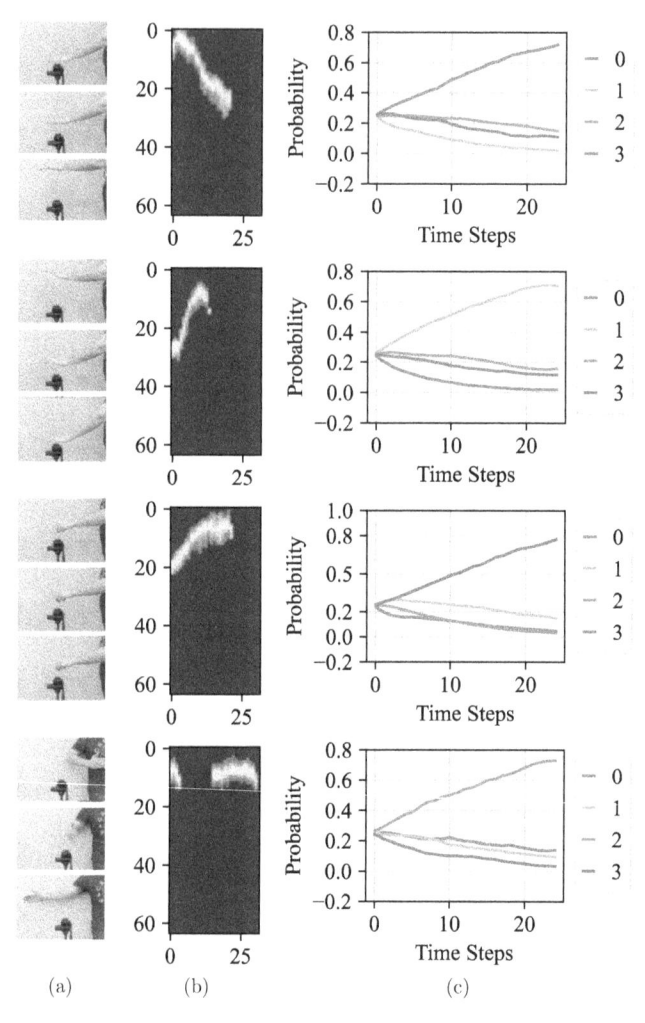

Fig. 3.12 Figure showing, (a) the four different gestures in the dataset, and the RD vector's map of four distinct gesture samples in (b), along with their firing patterns generated by the proposed SNN model in (c). The firing patterns demonstrate that the SNN begins to fire for a sample from the correct class after a few time steps due to presenting each image for a longer duration, resulting in improved accuracy

3.5 Optimized Gesture Sensing Using Range-Doppler Images

Fig. 3.13 Confusion matrix of the proposed SNN with the testing dataset, illustrating the confusion between down-up and finger rub gestures, and predominantly with right-left gestures due to similarities in their RD vector's maps

		Top-down	Down-up	Finger rub	Right-left
Actual Class	Top-down	100%	0	0	0
	Down-up	0	89.9%	1.77%	8.31%
	Finger rub	0	0	100%	0
	Right-left	0	0	0	100%

senting each image for a longer duration which allows for integrating spikes over a longer period of time, resulting in improved accuracy. Despite its simplicity, the proposed system attains several advantages over conventional deepNets, including improved energy efficiency, faster processing speed, and hardware compatibility while maintaining comparable levels of accuracy.

3.5.4 Conclusion

In this section, we presented an SNN-based gesture recognition system using 60-GHz FMCW radar. To incorporate the concept of time into spiking neural networks, we generated 2D range-Doppler vectors map based on range-Doppler images. This allowed us to capture temporal dynamics and enable temporal processing within the network. Compared to our previous approach, we present an innovative SNN classification architecture that eliminates the need for a convolutional layer, thereby meeting the specific hardware requirements. This was accomplished through meticulous refinement of the SNN architecture. The refinement of our signal processing chain ensures that the input data is appropriately preprocessed to extract meaningful features for the SNN classification architecture, removing the need for a convolutional layer. The proposed approach results in a similar level of classification accuracy as that of our proposed approach in Sect. 3.4 and state-of-the-art.

3.6 Gesture Sensing Using Range-Doppler and Angle Images

In this section, we propose an SNN-based gesture sensing system that can classify up to eight complex gestures. Additionally, we conducted a detailed assessment of various spiking neuron models, shedding light on their behavior using a complex dataset. The preprocessing pipeline is enhanced to generate more representative features such as range-time, velocity-time, and angle-time images, as features, surpassing the limitations of range-Doppler images, and facilitating enhanced learning within the predictive network. A modified and improved SNN architecture is proposed, achieving high accuracy in detecting and identifying eight different gestures. The content covered in this section corresponds to our publication P3.

3.6.1 Signal Processing Chain

3.6.1.1 Preprocessing
The preprocessing stage initiates by gathering raw ADC data throughout a chirp (fast-time) and organizing it into rows within the frame (slow-time). The ADC matrix undergoes various preprocessing techniques, including range FFT with coherent integration to enhance the signal-to-noise ratio. This is followed by the application of a moving target indicator to eliminate static targets within the field. Subsequently, target detection takes place, and the range bin corresponding to the target is selected.

3.6.1.2 Coherent Pulse Integration
In order to enhance the signal strength for target detection, we utilized pulse integration to merge the fast-time FFT data for each chirp in a frame, thereby improving the signal-to-noise ratio (SNR). Coherent pulse integration is utilized to coherently combine the phase and magnitude of the range FFT data across all chirps within a frame.

3.6.1.3 Moving Target Indication Filtering
In FMCW radar, the effectiveness of the initial FFT range bins is often hindered by TX-to-RX leakage. Moreover, reflections from the hand can be subdued by reflections from nearby stationary objects. To mitigate these challenges, we incorporated moving target indication (MTI) techniques to mitigate the impact of stationary objects and leakage. In each frame, a running average was applied to the

3.6 Gesture Sensing Using Range-Doppler and Angle Images

coherently integrated range FFT spectrum $R_{ci(k)}$ at the k-th frame. This operation can be mathematically represented as follows:

$$R_{ci}(k) = R_{ci}(k) - S(k-1) \tag{3.18}$$
$$S(k) = \alpha \cdot R_{ci}(k) + (1-\alpha) \cdot S(k-1) \tag{3.19}$$

where α denotes forget factor set to 0.01.The resulting filtered range FFT spectrum is then utilized as input for the subsequent target detection block.

Once the gesture initiation is detected, it is worth mentioning that the spectrograms can be generated without utilizing the MTI operation. Nonetheless, the high-pass filter property of the MTI operation is advantageous for enhancing the SNR by diminishing or eliminating any static hand reflection components present between consecutive frames in the spectrogram features. These components are irrelevant for the gesture classification application. The MTI operation highlights the hand movement or change, which is pivotal for accurate gesture classification.

Furthermore, the radar configuration used in this setup is designed to prioritize top-to-bottom illumination. This approach helps to minimize the presence of human body reflections or vital signals in the radar spectrograms. However, alternative configurations, such as mounting the radar near the web-camera of a laptop lid, may result in reflections from both the human body and vital signals being captured. In such a configuration, it becomes necessary to adapt the signal processing pipeline to rely on the first detected target within the range bin. This involves constructing the range, Doppler, and angle spectrograms based on the information obtained from

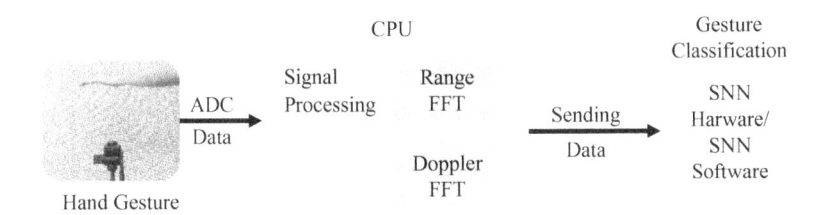

Fig. 3.14 The proposed system flow involves the raw data from the radar being initially input to a signal processing block, which carries out essential tasks like range FFT and Doppler FFT. The resulting data is then transferred to a gesture classification block that utilizes an SNN for classifying the gestures

the initial target detection. This may include applying a band-pass filter around the detected target to focus on its specific frequency range.

3.6.1.4 Target Detection

Target detection and selection are performed on the filtered range FFT spectrum using a straightforward thresholding technique. The threshold is determined by scaling the mean value of the range FFT spectrum. At frame k, the threshold Γ is calculated as follows:

$$\Gamma_k = \beta \times \sum_{n=1}^{N_s} R_{ci}^n(k), \tag{3.20}$$

Here, β is the scaling factor and is set to 3 specifically in our case. The index n denotes the range bins, and N_s represents the total number of range bins. The value of β was chosen empirically to achieve a balance between the probability of detection and the occurrence of false positives.

3.6.1.5 Range-Doppler Image

Generating the range-Doppler Image (RDI) involves mapping the reflected signal's frequency shift, induced by the target's range and velocity. The intensity of a pixel in the RDI represents the energy of the reflected signal, resulting in a blob-like structure. Creating a single RDI per frame involves applying a first-order FFT to the raw data along the fast time direction, transforming it into the range domain. Subsequently, a one-dimensional FFT is applied to the data in the range-Doppler domain. This process is reiterated for multiple frames, and the outcomes are combined into a three-dimensional cube, forming the RDI. Figure 3.15 visually outlines the entire process.

3.6.1.6 Angle of Arrival Estimation

The angle of arrival is a crucial parameter utilized for gesture recognition. In our approach, we have employed the minimum variance distortion-less response (MVDR) or Capon beam-former [106] to determine the angle or direction of arrival (DOA). For azimuth angle estimation, we have employed a Capon beam-former in the proposed system, which is expressed as a range-angle image (RAI) per frame.

3.6.1.7 Data Generation For the Model

Rather than using a computationally intensive 3D cube, we chose to simplify the representation of the Range-Doppler Image (RDI) by projecting it onto a 2D image. This was achieved by selecting the row with the highest pixel intensity in each frame and stacking these rows together, resulting in a 2D image known as a

3.6 Gesture Sensing Using Range-Doppler and Angle Images   77

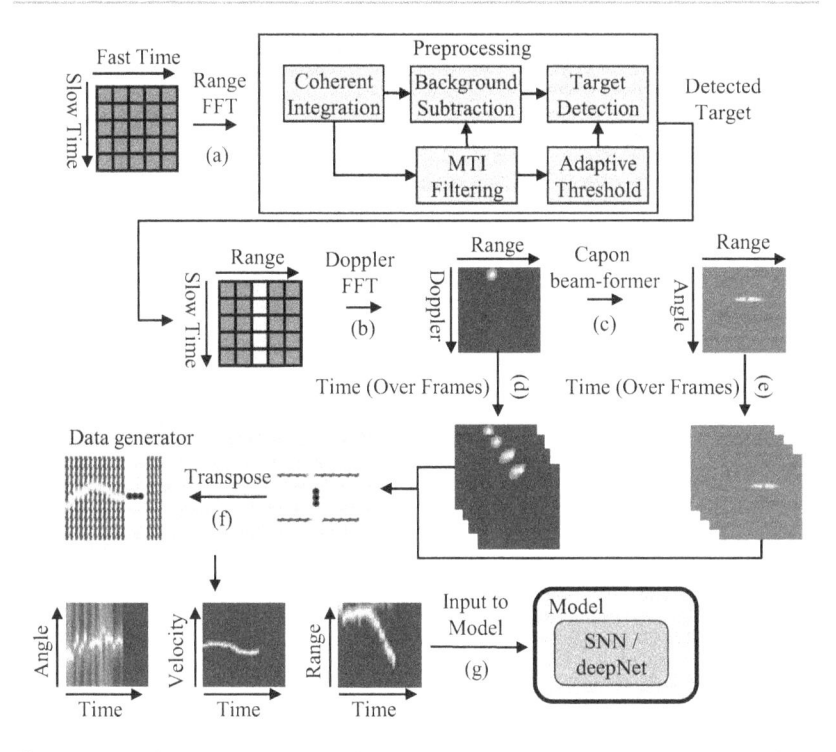

Fig. 3.15 The signal processing workflow for generating input to our proposed model: (a) performing preprocessing steps and applying 1D FFT along fast time, (b) conducting 1D FFT along slow time, (c) utilizing Capon beam-former to estimate the angle of arrival, (d) collecting and extracting range-Doppler image (RDI) over frames, (e) collecting and extracting range angle image (RAI) over frames, (f) stacking the RDI and RAI, and (g) using them as input to the model in the form of range-time, velocity-time, and angle-time images

range-time image. The same process was applied to generate velocity-time and angle-time images. Consequently, each gesture is characterized by three images: range-time, velocity-time, and angle-time images. This procedure is illustrated in Fig. 3.15.

3.6.2 Architecture & Learning

The primary aim of the proposed model is to maintain biological plausibility while minimizing computational requirements. To achieve this, the LIF neuron model was chosen. The LIF model was selected for its biologically plausible characteristics and its ability to reduce computational complexity.

3.6.2.1 Architecture

The proposed SNN architecture, implemented using NengoDL [38], is depicted in Fig. 3.16. The architecture begins with a $14 \times 14 \times 3$ input image, which undergoes processing through a convolutional layer featuring 16 filters of size 3 and a stride of 2. The output of the convolutional layer is then passed through LIF as a non-linear function, enabling the conversion of the output into spikes. Subsequently, the spike output is transmitted to a dense layer comprising 32 neurons, each of which is equipped with LIF. Finally, the network is equipped with a dense output layer that functions as the final layer. During training, the network employs SoftLIF [104] activation (a differential approximation to LIF) and utilizes a multi-class cross-entropy function as the objective.

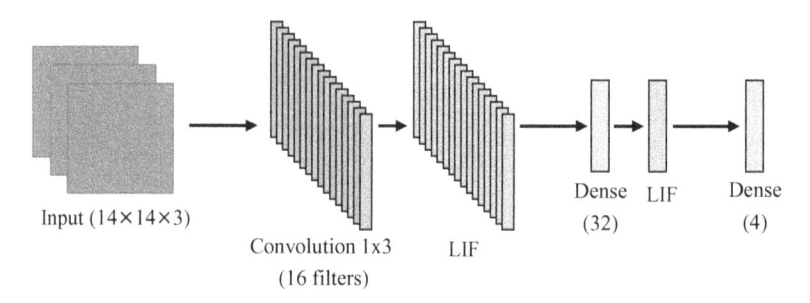

Fig. 3.16 The proposed SNN architecture with a $14 \times 14 \times 3$ input layer followed by a convolutional layer with 16 filters of size 3 and a stride of 2, appended with LIF, and a dense layer with 32 neurons appended with LIF. The network is trained with SoftLIF activation and a multi-class cross-entropy function, while in the testing phase, the trained network is reconstructed with LIF neurons. The weights and biases learned during training are used to connect spiking LIF neurons for testing, and the test inputs are presented multiple times or steps for accurate measurement of the spiking neuron output over time

3.6.2.2 Loss Function

To compute the classification probabilities, we utilized the softmax function for classification. The softmax function employs cross-entropy as the loss function for N training examples distributed among M classes. It is mathematically defined by the following equation:

3.6 Gesture Sensing Using Range-Doppler and Angle Images

$$L = -\frac{1}{N} \sum_{m=1}^{M} \sum_{n=1}^{N} y_n^m \log (h_\theta (x_n, m)), \tag{3.21}$$

where y_n^m are the true label for the n-th training example belonging to class m. The model h, which incorporates weights θ, takes input x for the given training example.

3.6.2.3 Learning Schedule and Weight Initialization

The Adam optimizer was utilized to optimize the learning rate for each parameter, employing adaptive learning rate techniques. During the training process, the learning rate (α) was set to 0.002, while the first (β_1) and second (β_2) moment estimates decayed exponentially with rates of 0.9 and 0.999, respectively. To maintain numerical stability, epsilon was set to $1e-8$. Dense layer weights were randomly drawn from a normal distribution with mean 0 and standard deviation 0.01, while biases were sampled from a normal distribution with mean 0.5. For convolutional layers, weights (w) were uniformly sampled within the range $[-\sqrt{\frac{6}{N_{in}+N_{out}}}, \sqrt{\frac{6}{N_{in}+N_{out}}}]$, where N_{in} and N_{out} are the number of input and output units, respectively.

3.6.2.4 Model Testing

During the testing phase, the trained network is reconstructed using LIF neurons, effectively transforming it into an SNN. The weights and biases acquired during the training process are extracted from the trained model and employed to establish connections among the spiking LIF neurons, constituting the testing model. To ensure precise measurements of the spiking neuron output over time, the test inputs or samples are repeatedly presented to the network during testing.

3.6.2.5 Model Hyperparameters

Hyperparameters are parameters that cannot be learned or derived from data, making them known as tuning parameters. These parameters play a critical role in determining the optimal performance of a model, as their values cannot be determined analytically. Identifying the optimal configurations of these hyperparameters enables models to learn more efficiently and achieve superior performance. The hyperparameters utilized for training the network are presented in Table 3.6. Furthermore, significant parameters for the LIF neuron, as outlined in [38], also play a crucial role:

1. τ_{rc}: the time constant for the membrane RC (resistance-capacitance) is measured in seconds and controls the rate at which the membrane voltage decays to zero in the absence of input.

Table 3.6 The hyperparameters utilized for the SNN model

Type	Parameters	Value
Hyperparameters	Learning rate (η)	0.001
	Output synaptic delay	10 ms
	Type of neuron model	LIF
	Batch size	100
	No. of epochs	10
	Time steps during the testing	25
	No. of layers	2
	No. of Filters	16

Table 3.7 The neuron parameters employed in the proposed SNN

Neuron Type	Parameters	Value
LIF	τ_{rc}	0.02
	τ_{ref}	0.002
	min voltage	0
	amplitude	0.01
Spiking LIF	τ_{rc}	0.02
	τ_{ref}	0.002
	amplitude	1
Spiking Tanh	τ_{ref}	0.0025
Spiking Sigmoid	τ_{ref}	0.0025
Spiking RectifiedLinear	amplitude	1

2. τ_{ref}: the absolute refractory period, expressed in seconds, sets the duration for which the membrane voltage remains at zero following a spike.
3. **Minimum voltage**: represents the minimum membrane voltage.
4. **Amplitude**: is the scaling factor by which the neuron output is scaled.

The values presented in Table 3.7 were selected for our experiments.

3.6 Gesture Sensing Using Range-Doppler and Angle Images

3.6.3 Results & Discussion

3.6.3.1 Dataset

In this proposed method we have used the complete dataset explained in Sect. 3.3.

3.6.3.2 Results

We used accuracy as the evaluation metric to assess the classification performance of our proposed system. The system's performance is comparable to that of state-of-the-art methods, as demonstrated by the similar average accuracy achieved over multiple trials, as shown in Table 3.8.

Table 3.8 shows the empirical comparison of the performance of various neuron models. Whereas, Fig. 3.19(d) demonstrates the firing choice prediction of the SNN model for some examples. Additionally, Fig. 3.17 depicts the trade-off between accuracy and input representation, while Fig. 3.18 displays the feature space visualization for system performance.

Table 3.8 The classification accuracy attained by the proposed SNN model in comparison to the existing models

Approaches	Accuracy	Model Size (kB)
CNN3D [107]	99.63%	12586.58
CNN2D [107]	86.25%	375.89
MobileNetV2 - 1 bottleneck [107]	98.88%	1770.96
MobileNetV2 - 2 bottleneck [107]	99.00%	2028.85
MobileNetV2 - 3 bottleneck [107]	97.13%	2287.06
MobileNetV2 - 4 bottleneck [107]	98.50%	2545.35
MobileNetV2 - 5 bottleneck [107]	97.75%	2804.27
MobileNetV2 - 6 bottleneck [107]	98.00%	3063.25
Custom model v_1 [107]	98.00%	624.92
Custom model v_2 [107]	97.50%	999.00
Custom model v_3 [107]	98.13%	1543.89
Custom model v_4 [107]	97.63%	2233.44
Proposed model_3	**99.50%**	**75**

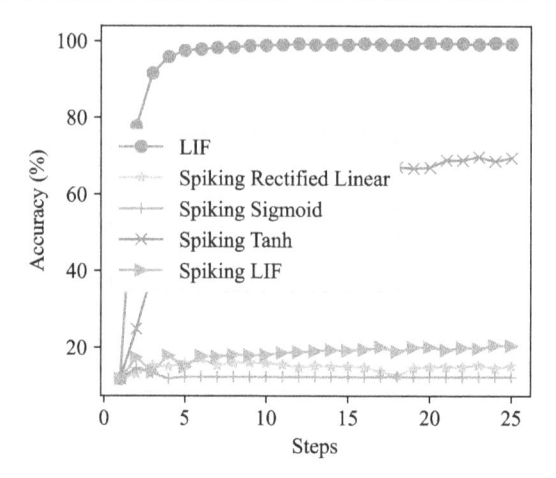

Fig. 3.17 The effect of presentation time on accuracy is shown, indicating that accuracy rises with an increase in the number of presentation steps until it reaches a plateau. The LIF model achieves peak accuracy with fewer steps compared to other models. To maintain consistency with state-of-the-art results for the LIF neuron, a presentation time of 25 steps was utilized for our comparisons

3.6.3.3 Discussion

We have demonstrated the utilization of SNN for gesture sensing through radar technology. A significant advantage of SNN is its energy efficiency attributed to its temporal processing capabilities. Our proposed gesture recognition system based on SNN is power-efficient and resource-friendly. In addition, SNNs are rapid, scalable, and straightforward to implement on hardware, making them an economical solution.

The firing pattern of the SNN for the eight gesture classes over time is illustrated in Fig. 3.19(d). It is noticeable that the SNN accurately fires for the provided sample image after a few time steps. This is attributed to the integration of spikes over an extended period, leading to higher accuracy in predicting the correct class as the time steps increase.

In our proposed SNN, we selected the LIF neuron model from various available options. This choice was made considering its biological plausibility and ease of implementation, as it requires fewer computational resources. To enhance the performance of the SNN, we conducted a grid search to fine-tune the network's hyperparameters. This approach allowed us to explore a range of hyperparameter combinations, enabling us to address both overfitting and underfitting scenarios.

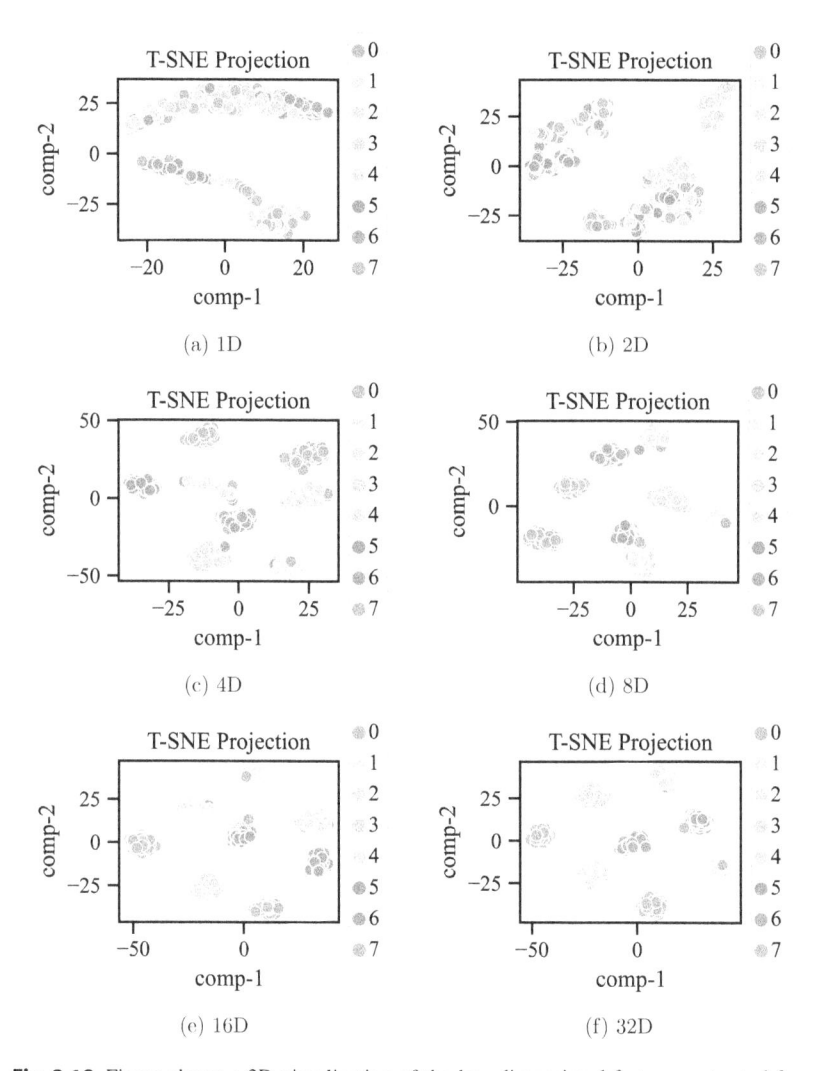

3.6 Gesture Sensing Using Range-Doppler and Angle Images

(a) 1D

(b) 2D

(c) 4D

(d) 8D

(e) 16D

(f) 32D

Fig. 3.18 Figure shows a 2D visualization of the low dimensional features extracted from the SNN for 8 different gestures using t-SNE. The plot demonstrates that the SNN has learned separable and discriminative features, with the 8 gesture classes forming tight clusters in just 4 dimensions of embedding space. This suggests that the SNN is capable of reliably classifying the 8 gesture classes even in low-dimensional feature space

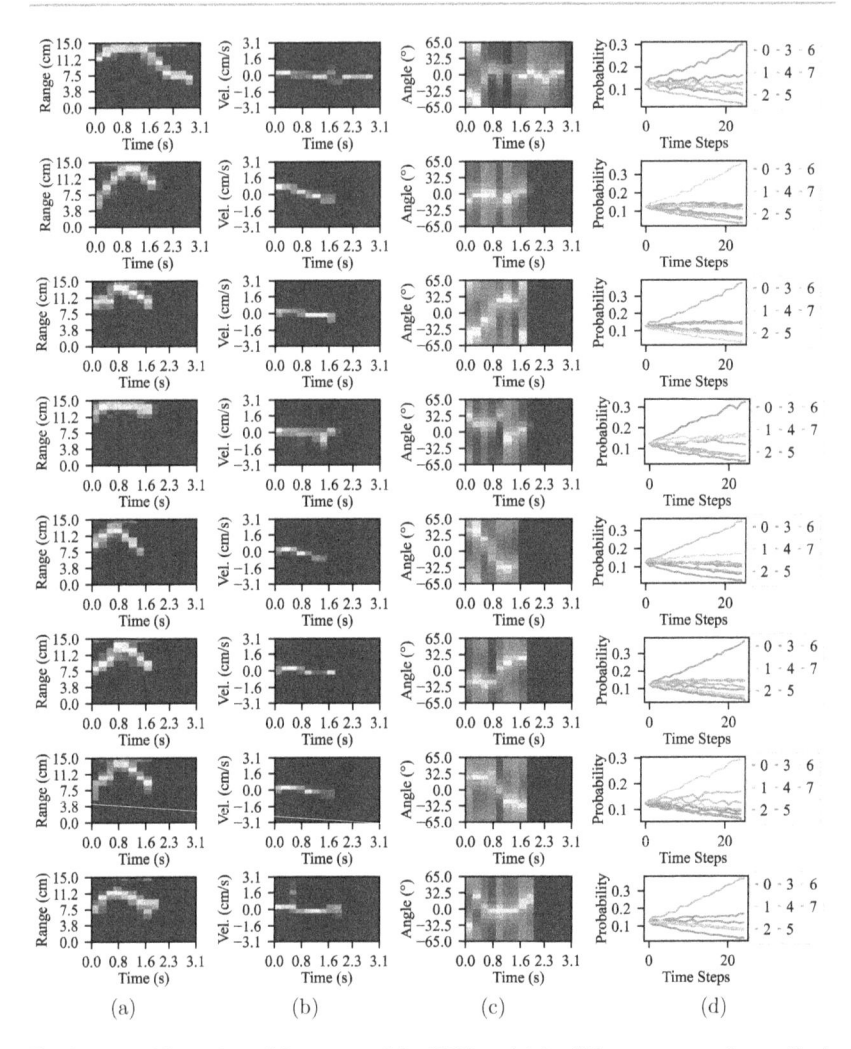

Fig. 3.19 An illustration of the output of the SNN model for different gesture classes. Each row represents a sample from one of the eight gesture classes, and the columns (a), (b), and (c) show the range, velocity, and angle spectrogram, respectively. The firing choice of the SNN model for each sample is displayed in column (d). Starting from top row the eight gesture classes are 0—down up, 1—up down, 2—left-right, 3—rubbing, 4—right-left, 5—diagonal southwest to northeast, 6—diagonal southeast to northwest, and 7—clapping

3.6 Gesture Sensing Using Range-Doppler and Angle Images

Table 3.6 demonstrates that the SNN performs better with specific hyperparameter settings. Additionally, we observed that the neuron parameter configurations illustrated in Table 3.7 contribute to improving the SNN's performance at the neuron level.

Utilizing the parameters outlined in Table 3.6 and Table 3.7, our proposed system can accurately classify eight different gestures with an accuracy (99.50%) level similar to conventional deepNets, as demonstrated in Table 3.8.

By employing the parameters specified in Table 3.6 and Table 3.7, our proposed system achieves a high level of accuracy (99.50%) in accurately classifying eight distinct gestures. This accuracy level is comparable to conventional deepNets, as indicated in Table 3.8. The proposed system's exceptional performance can be attributed to the spatio-temporal information encoding by the SNN, leveraging the network dynamics for learning. Furthermore, with a size of only 75kB, the proposed model is significantly smaller than its deepNet counterpart, making it more memory efficient.

To evaluate and compare the performance of different neuron types, training neurons, neuron models, and optimization techniques, we conducted an empirical analysis using an identical network architecture. Our objective was to identify the strengths of each approach in terms of accuracy, simulation time, and training time across equivalent epoch runs.

In our investigation, we considered diverse neuron models, including LIF and spiking variants of rate-based neurons, such as LIF Rate, Rectified Linear, Sigmoid, and Tanh. These models were evaluated using two optimization techniques: RMSProp and Adam. By employing this comprehensive approach, we aimed to gain insights into the performance characteristics of each model and optimization method. The results of our empirical study, presented in Table 3.9, demonstrate that the LIF neuron type utilizing Soft LIF achieves outstanding classification accuracy. Furthermore, the simulation and training times for this configuration are on par with those of other neuron types and training neurons. Importantly, it should be noted that for the LIF neuron type and training neuron, both optimizers produce similar accuracy results, which is not the case for other neuron types or training neurons.

One of crucial element that impacts the accuracy is the frequency at which we present input to the network. Given that we are dealing with static images, it's important to select an appropriate presentation time to ensure that the network has enough time to accumulate the necessary current to make accurate predictions or firings for each class. When considering presentation time, there is a trade-off between accuracy and latency. Longer presentation times enable the integration of spikes over a greater time period, leading to improved accuracy. Conversely, shorter presentation times yield higher throughput and lower latency at the expense of some

Table 3.9 Comparing the classification accuracy of the top-performing models to achieve improved accuracy

Neuron Type	Accuracy	Optimizer	Training Neuron	Simulation Time	Training Time	No. of Epochs
LIF	**99.37**	**RMSprop**	**Soft LIF**	**0.52**	**9.29**	**10**
	99.50	**Adam**	**Soft LIF**	**0.5**	**8.54**	**10**
Regular Spike Interval						
Spiking LIF	22.0	RMSprop	LIF Rate	0.5	12.58	10
	21.5	Adam	LIF Rate	0.49	12.03	10
Spiking Rectified Linear	24.5	RMSprop	Rectified Linear	0.48	10.7	10
	17.75	Adam	Rectified Linear	0.5	10.55	10
Spiking Sigmoid	16.75	RMSprop	Sigmoid	0.47	10.78	10
	16.88	Adam	Sigmoid	0.8	10.5	10
Spiking Tanh	56.5	RMSprop	Tanh	0.46	11.07	10
	29.62	Adam	Tanh	0.5	10.69	10
Stochastic Spiking						
Spiking LIF	19.37	RMSprop	LIF Rate	0.5	13.37	10
	14.62	Adam	LIF Rate	0.49	12.75	10
Spiking Rectified Linear	25.5	RMSprop	Rectified Linear	0.48	11.31	10
	15.38	Adam	Rectified Linear	0.48	11.13	10
Spiking Sigmoid	15.25	RMSprop	Sigmoid	0.49	11.09	10
	10.63	Adam	Sigmoid	0.9	10.88	10
Spiking Tanh	42.25	RMSprop	Tanh	0.49	11.39	10
	20.13	Adam	Tanh	0.51	10.85	10
Poisson Spiking						
Spiking LIF	18.25	RMSprop	LIF Rate	0.5	12.04	10
	14.5	Adam	LIF Rate	0.49	12.05	10
Spiking Rectified Linear	23.0	RMSprop	Rectified Linear	0.49	10.74	10
	14.5	Adam	Rectified Linear	0.47	10.43	10
Spiking Sigmoid	14.0	RMSprop	Sigmoid	0.46	11.4	10
	14.88	Adam	Sigmoid	0.47	10.61	10
Spiking Tanh	34.62	RMSprop	Tanh	0.52	10.93	10
	17.5	Adam	Tanh	0.46	10.73	10

accuracy. The impact of presentation time on accuracy is depicted in Fig. 3.17. As the number of presentation steps increases, accuracy also rises until it reaches a plateau. Notably, compared to other models, the LIF model attains peak accuracy with fewer steps. To facilitate our comparisons and achieve results comparable to state-of-the-art for the LIF neuron, we opted for a presentation time of 25 steps.

To further investigate our system's performance, we visualized the high-dimensional feature space. The convolution layers extract features from the input, and we examined how effectively these features discriminate between classes by inputting them into the t-Distributed Stochastic Neighbor Embedding algorithm (t-SNE) alongside their corresponding labels. Similar to principal component analysis, t-SNE is a popular method for reducing the dimensionality of high-dimensional

3.6 Gesture Sensing Using Range-Doppler and Angle Images

datasets for visualization purposes. However, unlike PCA, t-SNE is a probabilistic technique that measures pairwise neighbor similarities in both the high-dimensional and corresponding low-dimensional spaces using the ℓ_2 norm metric. A 2D visualization of the low-dimensional features for 8 gestures using t-SNE can be seen in Fig. 3.18. The plot demonstrates that our SNN network has learned both separable and discriminative features, as evidenced by the tight clusters formed by the 8 gesture classes in just 4 dimensions of embedding space. This illustrates that even a low-dimensional SNN can reliably classify the 8 gesture classes.

To assess the energy efficiency of our proposed system, we evaluated its energy consumption per classification. To do so, we utilized the hardware metrics of the μBrain chip outlined in [108], which can be expressed mathematically as:

$$E_c = N_{spikes} \times E_{spikes} + \delta T \times P_{leakage}, \tag{3.22}$$

where E_c denotes the energy consumed per classification, N_{spikes} represents the maximum number of spikes generated during classification, $E_{spikes} = 2.1\,\text{pJ}$ is the energy per spike, $P_{leakage} = 73\,\mu\text{W}$ denotes the static leakage power, and δT represents the inference time. Assuming an inference time of $\delta T = 28\,\text{ms}$, the energy consumption per classification of our proposed system is $E_c = 2.04\,\mu\text{J}$.

The proposed SNN system is characterized by its simplicity and effectiveness, demonstrating high accuracy in real-time hand gesture recognition with low latency, comparable to deepNets counterparts. Furthermore, SNNs exhibit energy efficiency by consuming power only when their corresponding neurons fire, which is inherently sparse. This feature positions them as an ideal solution for low-power hardware implementation, making them well-suited for consumer applications.

3.6.4 Conclusion

This section introduces a gesture recognition system based on a spiking neural network, utilizing FMCW radar. To extract relevant features, we propose the utilization of range spectrograms, Doppler spectrograms, and angle spectrograms derived from video data of range-Doppler images. The proposed spiking neural network architecture is particularly suitable for low latency and low power embedded implementations, making it an appealing choice for human-machine interface applications. A critical aspect of training spiking neural networks lies in optimizing the learning rules to effectively capture spatiotemporal spike patterns. Our results demonstrate that our proposed system, featuring optimized learning rules and compact dimensions, achieves classification accuracy comparable to deepNet models across eight distinct gestures.

3.7 Gesture Recognition System Using Raw ADC Data

In this section we propose an embedded gesture recognition system that utilizes a 60-GHz FMCW radar, employing spiking SNNs directly on the raw ADC data. Unlike previous state-of-the-art methods, our system solely relies on the raw ADC data, eliminating the need for time-consuming slow-time and fast-time FFTs. Additionally, our proposed SNN architecture mimics the pre-processing slow-time FFT, achieving a processing speed of 112ms, surpassing existing methods by more than 2 times compared to our proposed architecture in Sects. 3.4 and 3.5. Experimental results validate the effectiveness of our approach, demonstrating a recognition accuracy of 98.1%. This performance is comparable to our previous approaches Sects. 3.4 and 3.5 and state-of-the-art approaches, despite the simplification of our implementation. The content covered in this section corresponds to our publication P2.

3.7.1 Signal Processing Chain

3.7.1.1 Moving Target Indication Filtering

The raw ADC data is gathered during a chirp (fast-time) and organized into rows along with the frame (slow-time). When using FMCW radar, stationary object reflections in the environment can overshadow the object reflections of interest. Therefore, we utilize MTI to suppress the stationary object reflections and leakage. During each frame j, we apply a moving average filter to the fast-time $S(j)$, expressed mathematically as follows:

$$S(j) = S(j) - K(j-1), \qquad (3.23)$$
$$K(j) = \alpha \cdot S(j) + (1-\alpha) \cdot K(j-1), \qquad (3.24)$$

where the forget factor α is set to 0.01. $K(j)$ denotes the average weight of the current fast-time $S(j)$ and the previous MTI value $K(j-1)$. We set the initial value of $K(j)$ to 0. The target detection block receives the filtered fast-time data.

3.7 Gesture Recognition System Using Raw ADC Data

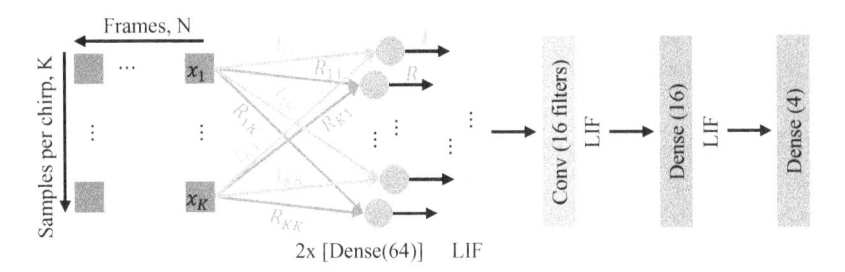

Fig. 3.20 The proposed SNN architecture's initial layer mimics the behavior of the discrete Fourier transform (DFT). This layer consists of $2 \times N_S$ nodes, calculating real and imaginary values using weights determined by DFT trigonometric Eq. 3.25. The input data dimensions are $N_S \times N_F$, where N_S refers to total samples per chirp and N_F denotes total frames. Variables k and l range from 0 to $N_S - 1$. After this layer, a convolutional layer with 16 filters, size 3, and stride 2 follows. Its output undergoes LIF-based non-linear processing and enters a fully connected layer with 16 neurons also utilizing LIF for activation. The final model layer, with 4 neurons, handles classification

3.7.2 Architecture & Learning

The proposed SNN architecture, depicted in Fig. 3.20, focuses on resource efficiency in terms of computation and energy. To achieve this, we adopt LIF neurons in our design. The model is constructed using NengoDL [38], which enables a differential approximation of the firing rate of LIF neurons and facilitates training through conventional backpropagation. For activation, we employ SoftLIF [104] (an approximation to LIF), and the objective function is defined as multi-class cross-entropy.

The first layer of the SNN is designed to simulate the functionality of the Discrete Fourier Transform (DFT). Since the DFT is a linear transformation that can be decomposed into two consecutive multiplications, we can model each dimension of the DFT using a single dense layer. Assuming the input data has dimensions of $N_S \times N_F$, where N_S represents the total number of samples per chirp and N_F represents the total number of frames, the layer is configured with $2 \times N_S$ nodes to compute both the real and imaginary values. The weights of the layer are determined based on the DFT trigonometric equation:

$$D_k = \sum_{l=0}^{L-1} X_l \left[\cos\left(\frac{2\pi}{L}kl\right) - i \sin\left(\frac{2\pi}{L}kl\right) \right], \quad (3.25)$$

The variables k and l takes value from 0 to $N_S - 1$. Equation (3.25) can be expressed in matrix form as follows:

$$D = (W_R + iW_I) X, \qquad (3.26)$$

where D represents the outcome of the transformation, X denotes the input vector, and W_R and W_I correspond to the real and imaginary coefficients. Following the initial layer, the generated output is fed into a convolutional layer, utilizing a filter size of 3, incorporating a total of 16 filters, and applying a stride of 2. Subsequently, the output of this layer is processed by LIF as a non-linear function, effectively transforming the output into spike representations. Next, a fully connected layer with 16 neurons is employed, utilizing LIF as the activation function. Finally, a classification layer consisting of 4 neurons is utilized as the output of the model. During the testing phase, the softLIF activation function is replaced with LIF, enabling the transformation of the network into a spiking neural network. The weights and biases acquired from the trained model are subsequently utilized to establish connections between LIF neurons. For precise spiking neuron behavior measurements, test inputs are repeatedly presented to the network.

3.7.3 Results & Discussion

3.7.3.1 Dataset

Four gestures were chosen from the dataset described in Sect. 3.3. These gestures include: 1) top-down (moving hand downward), 2) down-up (moving hand upward), 3) right-left (moving hand to the left), and 4) rubbing two fingers together.

3.7.3.2 Results

The effectiveness of the proposed system is evaluated by measuring its classification accuracy. The system achieves an average accuracy of 98.1%, which is consistent with state-of-the-art models such as Proposed model_1 and Proposed model_2 proposed in Sect. 3.4 and Sect. 3.5, respectively. Furthermore, the proposed system's performance is comparable to that of deep neural networks such as Long short-term memory (LSTM), 1D Convolution LSTM (1DCNN-LSTM), 2D Convolution LSTM (2DCNN-LSTM), and Temporal Convolutional Networks (TCN) across multiple random tests, as shown in Table 3.10. Moreover, it is evident that our proposed model has notably lower latency in comparison to the state-of-the-art (Proposed model_1 and Proposed model_2 in Sect. 3.4 and Sect. 3.5 respectively) and equivalent deep neural networks.

3.7 Gesture Recognition System Using Raw ADC Data

Table 3.10 Comparing classification accuracy: proposed SNN model versus other SNN and conventional ANN models

Approaches	Input type	Input Size	Accuracy	Latency (ms)
LSTM	Range over time	64 × 32	96.9%	311
2DCNN-LSTM	Range over time image	64 × 32	97.6%	290
TCN	Range over time	64 × 32	98.9%	960
1DCNN-LSTM	Range over time	64 × 32	97.2%	315
Proposed model_1	Range over time image	64 × 32	98.5%	258
Proposed model_2	Range over time image	64 × 32	97.5%	229
Proposed model_4	**ADC**	**64 × 32**	**98.1%**	**112**

3.7.3.3 Discussion

In this section, we present a gesture sensing system that utilizes SNNs. Unlike conventional deep neural networks, SNNs take into account both the static values and the timing of these values [105]. This approach offers increased energy efficiency by employing fewer neurons. Moreover, SNNs demonstrate speed, scalability, and compatibility with hardware, making them a cost-effective solution.

The proposed solution stands out from previous SNN approaches as it relies solely on the raw data of the target, eliminating the need for slow-time and fast-time FFT operations. Furthermore, unlike previous methods that utilize range-over-time images of dimensions 64 × 32 with 32 chirps per frame, our system only requires a single chirp per frame. This reduction in chirps significantly reduces the computational cost while maintaining a comparable level of accuracy in classifying four gestures, as shown in Table 3.10. The performance of the proposed system can be attributed to its utilization of spatiotemporal information encoding through SNNs, which effectively exploits network dynamics for the learning process. Moreover, our proposed SNN demonstrates an inference latency that is more than 2× fast compared to the state-of-the-art SNN methods. In this context, latency refers to the time required for the system to perform inference following the execution of a gesture.

Figure 3.21 illustrates examples of four distinct gesture samples and their corresponding firing selection by the model. The figure depicts the SNN's firing pattern, where, after a few time steps, the neurons begin to fire accurately for samples belonging to the correct classes. This observation can be attributed to the prolonged presentation of each image, enabling the integration of spikes over an extended duration and resulting in higher accuracy.

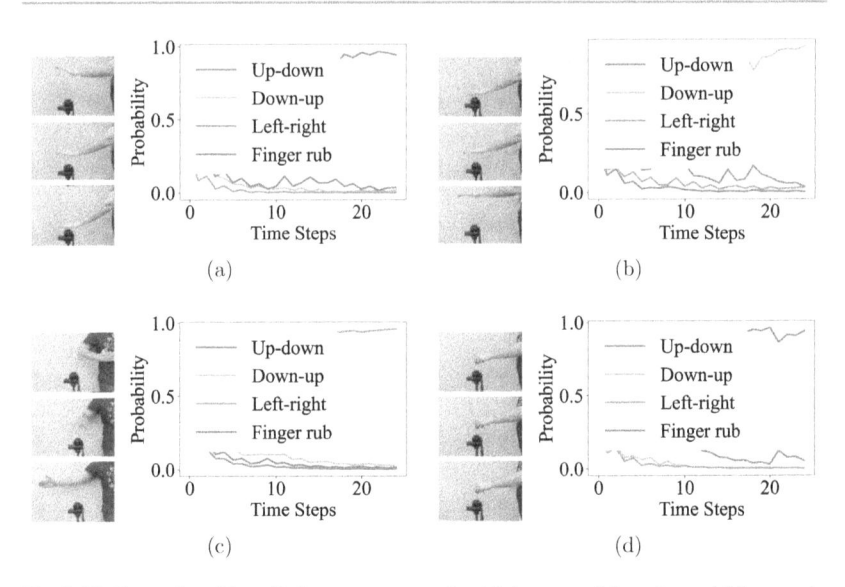

Fig. 3.21 Examples of four distinct gesture samples (a) down-up, (b) up-down, (c) finger rub, and (d) left-right, along with the model's chosen firing patterns. It becomes evident that the SNN starts firing for the correct class samples within a short time, showcasing its ability to classify accurately. The enhanced accuracy stems from prolonging the image's presentation time, enabling spike integration over a more extended period

To assess its energy efficiency, we analyzed the energy consumption per classification. We utilized the hardware metrics of the μBrain chip, which are defined in [108] and expressed mathematically as follows:

$$E_c = N_{\text{spikes}} \times E_{\text{spikes}} + \delta T \times P_{\text{leakage}} \quad (3.27)$$

where E_c denotes the energy consumed per classification, N_{spikes} represents the maximum number of spikes during classification, $E_{\text{spikes}} = 2.1\,\text{pJ}$ signifies the energy per spike, $P_{\text{leakage}} = 73\,\mu\text{W}$ represents the static leakage power, and δT denotes the inference time. Assuming δT to be 28 ms, the proposed system achieves an energy consumption per classification of $E_c = 2.05\,\mu\text{J}$.

3.7.4 Conclusion

In this section, we introduce an innovative gesture sensing system based on SNNs and 60-GHz FMCW radar. Unlike existing approaches that rely on Doppler images, our method operates solely on raw ADC data. This eliminates the need for time-consuming slow-time and fast-time FFTs, resulting in a more than $2\times$ improvement in end-to-end processing speed. The evaluation of our proposed SNN architecture on four distinct gestures demonstrates comparable accuracy performance to state-of-the-art SNNs. The low power consumption of the proposed system makes it suitable for embedded implementations.

3.8 Mimicking Fourier Transforms with Spiking Neural Networks

In this section, we present an embedded gesture detection system that directly utilizes SNNs on the raw ADC data obtained from a 60-GHz FMCW radar. Our proposed system stands out from previous state-of-the-art methods as it solely relies on the raw ADC data of the target, eliminating the need for computationally expensive slow-time and fast-time FFT processing. To achieve this, we mimic the discrete Fourier transformation within the SNN architecture itself, removing the reliance on FFT accelerators and tailoring the processing specifically for gesture sensing applications. Experimental results showcase the impressive performance of our proposed system, achieving an accuracy of 98.7% in accurately classifying eight different gestures. This level of accuracy is comparable to conventional approaches while offering the additional advantages of lower complexity, reduced power consumption, and faster computations compared to traditional methods. The content presented in this section aligns with our publication P1.

3.8.1 Signal Processing Chain

Figure 3.22 illustrates the proposed signal processing chain employed in the proposed approach. The process begins with the CPU receiving the raw ADC data from the radar. Subsequently, the CPU performs crucial signal processing tasks, such as the implementation of MTI and target detection algorithms. The filtered data is then forwarded to specialized hardware or software specifically designed for SNN-based classification tasks.

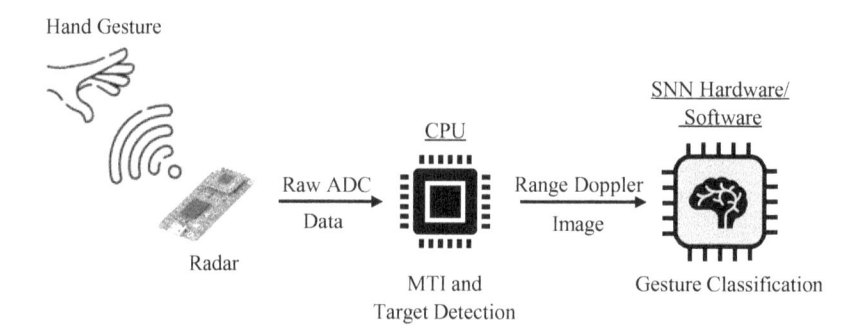

Fig. 3.22 The proposed processing chain involves connecting the radar to a CPU. The CPU receives raw ADC data and performs signal processing tasks, including MTI and target detection. The filtered data is then forwarded to SNN hardware or software for gesture classification

3.8.1.1 Raw Data

The raw ADC data, denoted as $I_F(t)$, is collected from the radar chipset in a chirp-wise manner within the fast-time domain. These chirps are then stacked together in rows, forming a representation of the data in the slow-time domain. Consequently, each frame is represented by a 2D array, where each row corresponds to a specific chirp.

3.8.1.2 Moving Target Indication Filtering

The reflections from stationary objects in radar signals can become stronger than the reflection from the intended target (in this instance, a hand), leading to interference. To mitigate this issue, we employed a technique called MTI. MTI involves applying a moving average filter to the fast-time data at each frame (i) to suppress reflections from stationary objects. The filter is defined mathematically as follows:

$$S(j) = S(j) - K(j-1), \qquad (3.28)$$
$$K(j) = \alpha \cdot S(j) + (1-\alpha) \cdot K(j-1), \qquad (3.29)$$

where the forget factor α is assigned a value of 0.01. $K(j)$ represents the average weight between the current fast-time data $S(j)$ and the previous MTI value $K(j-1)$. The initial value of $K(j)$ is set to 0. The target detection block then takes in the filtered fast-time data.

3.8.2 Architecture & Learning

The proposed architecture aims to be computationally and energy efficient. To achieve this goal, we have selected the LIF neuron model. Figure 3.23 illustrates the proposed SNN architecture. However, since the LIF is not differentiable, backpropagation is not possible. To address this limitation, we utilized a differentiable approximation of the LIF.

In order to emulate the DFT within SNN layers, we leverage the linear transformation property of the DFT's successive multiplication representation. To achieve this, we use a single Dense layer to represent each DFT dimension, where the layer weights correspond to the real and complex components of the DFT coefficients. Consider a radar system with a total of F_n frames and S_n samples per chirp, resulting in an input data dimension of $S_n \times F_n$. In order to perform the Discrete Fourier Transform (DFT) on this input data, the first layer of our model comprises $2 \times S_n$ nodes, responsible for computing the real and imaginary values. The connectivity between the input nodes and the layer nodes consists of $2S_n \times S_n$ connections. To determine the connection weights, we utilize the DFT trigonometric equation, which can be expressed as follows:

$$C_q = \sum_{p=0}^{P-1} Y_p \left[\cos\left(\frac{2\pi}{P}qp\right) - i \sin\left(\frac{2\pi}{P}qp\right) \right], \tag{3.30}$$

where q and p are indices that vary from 0 to $S_n - 1$. When this equation is applied to an input vector Y, it can be represented in matrix form as:

$$C = (W_R + i W_I) Y, \tag{3.31}$$

where C represents the resulting transform, and W_R and W_I denote the real and imaginary coefficients, respectively.

As the second FFT is applied across slow-time in radar processing, the output of the first layer is reshaped and transposed using a transpose layer. Then, in the third layer, the real and imaginary parts of the output from the first layer are connected separately to the real and imaginary weights calculated using the trigonometric equation mentioned earlier. Let Y^r represents the transformation $Y^T W_R$ and Y^i represents the transformation $Y^T W_I$ then at layer 3 following transformation occurs:

$$C^r = \left(W_R^r + i W_I^r\right) Y^r, \tag{3.32}$$

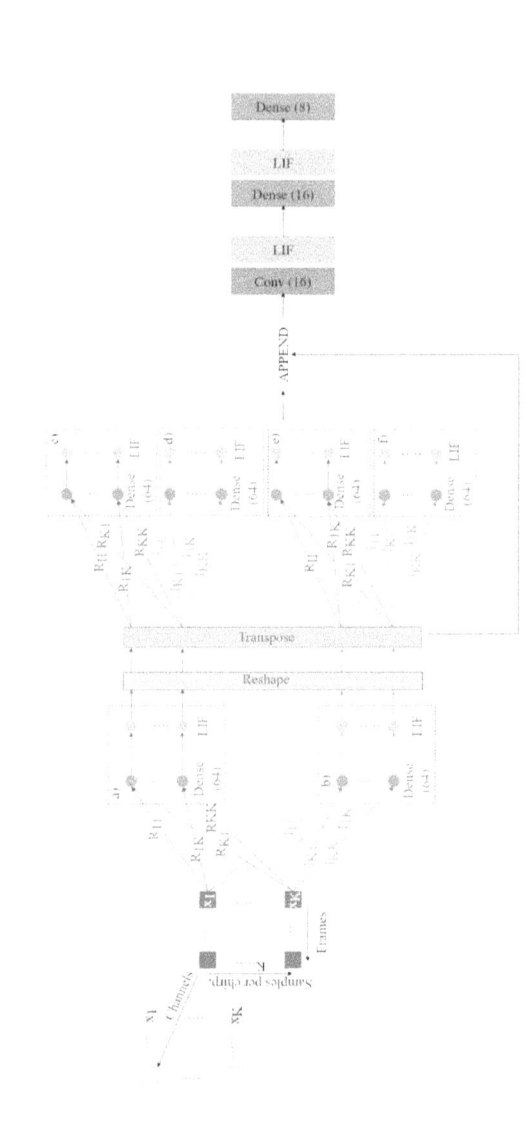

Fig. 3.23 The proposed SNN architecture emulates the characteristics of slow-time and fast-time FFTs in its initial layers, followed by the integration of a CNN and dense layer for classification purposes. Within this architecture: In block a) the input is multiplied with the real coefficients of the FFT, resulting in Y^r. In block b) the input is multiplied with the imaginary coefficients of the FFT, yielding Y^i. These operations mimic the range FFT. Similarly, block c) involves the multiplication of Y^r with the real coefficients of the FFT, leading to Y^{rr} and block d) represents the multiplication of Y^r with the imaginary coefficients of the FFT, leading to Y^{ri}. Likewise, block e) executes the multiplication of Y^i with the real coefficients of the FFT, generating Y^{ri}. Block f) represents the multiplication of Y^i with the imaginary coefficients of the FFT, resulting in Y^{ii}. Subsequently, the outputs Y^r, Y^i, Y^{rr}, Y^{ri}, Y^{ir}, and Y^{ii} are appended and fed to the convolution layer, followed by the dense and output layers

3.8 Mimicking Fourier Transforms with Spiking Neural Networks

$$C^i = \left(W_R^i + i W_I^i\right) Y^i. \tag{3.33}$$

Both the first and third layers are equipped with LIF as their activation function, which converts the output of each neuron into spikes. Let us denote the output of the third layer as:

$$Y^{rr} = Y^{r^T} W_R^r, \tag{3.34}$$

$$Y^{ri} = Y^{r^T} W_I^r, \tag{3.35}$$

$$Y^{ir} = Y^{i^T} W_R^r, \tag{3.36}$$

$$Y^{ii} = Y^{i^T} W_I^i. \tag{3.37}$$

Next, the outputs of the first and third layers are concatenated with each other, resulting in:

$$\rho = [Y^r, Y^i, Y^{rr}, Y^{ri}, Y^{ir}, Y^{ii}]. \tag{3.38}$$

The concatenated output is then passed through the convolutional layer, where each transformation corresponds to a channel. Consequently, ρ is transformed into a six-dimensional vector. The convolutional layer comprises of 16 filters, each with a size of 3 and a stride of 1. The resulting output is then forwarded to a fully connected layer with 64 neurons, employing LIF as the activation function. LIF is also used as the activation function in both the convolutional and fully connected layers. Finally, a fully connected layer with 8 neurons generates the output of the model.

3.8.2.1 Training

For network training, we employ NengoDL [38], which facilitates the use of conventional backpropagation. This is achieved by leveraging NengoDL's ability to differentially approximate the firing rate of LIF neurons through SoftLIF [104] activation. Multi-class cross-entropy is employed as the objective function to compute classification probabilities.

Similar to ANNs, the learning performance of SNNs depends on appropriate optimization solvers and weight initialization. In our study, we utilized the adaptive moment estimation (Adam) [109] as the optimization solver, which has been shown to be a computationally efficient and effective candidate for large networks [110].

We use a softmax classifier as the loss function, with cross-entropy as the associated loss metric. Mathematically, the cross-entropy for J training samples belonging to K classes can be expressed as:

$$E_C = -\frac{1}{M} \sum_{q=1}^{Q} \sum_{m=1}^{M} z_m^q \log\left(H_\theta\left(x_j, q\right)\right), \tag{3.39}$$

where z_m^q represents the true label for the mth training example of class q, and x denotes the input example to the model H with weights θ.

3.8.2.2 Testing

To transform the network into a spiking one, we substitute the LIF neurons in the trained model with spiking LIF neurons. The connection weights and neuron biases required for the spiking LIF neurons are derived from the trained model. To ensure precise measurements of the spiking neuron output over time, we adjust the test inputs or samples during testing by presenting them to the network multiple times or steps.

3.8.3 Results & Discussion

3.8.3.1 Dataset
In this proposed method we have used the complete dataset explained in Sect. 3.3.

3.8.3.2 Results
To assess the effectiveness of our proposed system, we utilized classification accuracy as the performance metric. The average accuracy achieved by our proposed system was 98.7%, which is comparable to the accuracy achieved by state-of-the-art methods across multiple random trials, as presented in Table 3.11. All methods were applied to the same dataset. In our proposed models (Proposed model_1, Proposed model_2, and Proposed model_4 in Sects. 3.4, 3.5 and 3.7 respectively), we employed four out of the eight gestures from the same dataset for evaluation.

3.8.3.3 Discussion
This section presents a novel radar-based gesture sensing system that leverages SNNs to eliminate the need for traditional radar signal pre-processing steps, such as fast-time and slow-time FFTs followed by Constant False Alarm Rate (CFAR). The proposed system performs gesture sensing directly on raw ADC data, simulating the fast-time and slow-time FFTs within the SNN architecture. Additionally, the SNN learns the CFAR and other denoising steps, showcasing the comprehensive capabilities of SNNs. By harnessing the efficiency, speed, scalability, and hardware-friendly nature of SNNs, the proposed system offers a cost-effective and energy-efficient solution for gesture sensing applications.

3.8 Mimicking Fourier Transforms with Spiking Neural Networks

Table 3.11 Comparison of the classification accuracy of the proposed models with other existing models, including RVA (range over time images, velocity over time images, and angle over time images) and ROT (range over time image) models

Approaches	Total gestures	Input type	Accuracy	Model size (kB)
CNN3D [107]	8	RVA	99.63%	12586.58
CNN2D [107]	8	RVA	86.25%	375.89
MobileNetV2 - 1 bottleneck [107]	8	RVA	98.88%	1770.96
MobileNetV2 - 2 bottleneck [107]	8	RVA	99.00%	2028.85
MobileNetV2 - 3 bottleneck [107]	8	RVA	97.13%	2287.06
MobileNetV2 - 4 bottleneck [107]	8	RVA	98.50%	2545.35
MobileNetV2 - 5 bottleneck [107]	8	RVA	97.75%	2804.27
MobileNetV2 - 6 bottleneck [107]	8	RVA	98.00%	3063.25
Custom model v_1 [107]	8	RVA	98.00%	624.92
Custom model v_2 [107]	8	RVA	97.50%	999.00
Custom model v_3 [107]	8	RVA	98.13%	1543.89
Custom model v_4 [107]	8	RVA	97.63%	2233.44
Proposed model_1	4	ROT	98.5%	257
Proposed model_2	4	ROT	97.5%	41
Proposed model_3	8	RVA	99.50%	75
Proposed model_4	4	Raw ADC	98.1%	515
Proposed model_5	8	Raw ADC	**98.7%**	14824

The proposed method not only circumvents the need for traditional feature engineering techniques by incorporating slow-time and fast-time FFT operations within the SNN network but also deviates from existing state-of-the-art methods that rely on utilizing 32 chirps per frame. Instead, the proposed approach operates with a single chirp per frame. This significant reduction in the number of chirps per frame leads to a substantial decrease in computational costs, making the proposed method more computationally efficient compared to previous approaches.

Figure 3.24 illustrates the firing patterns and predictions of the proposed system over time for various examples. Initially, at the beginning of the observation period,

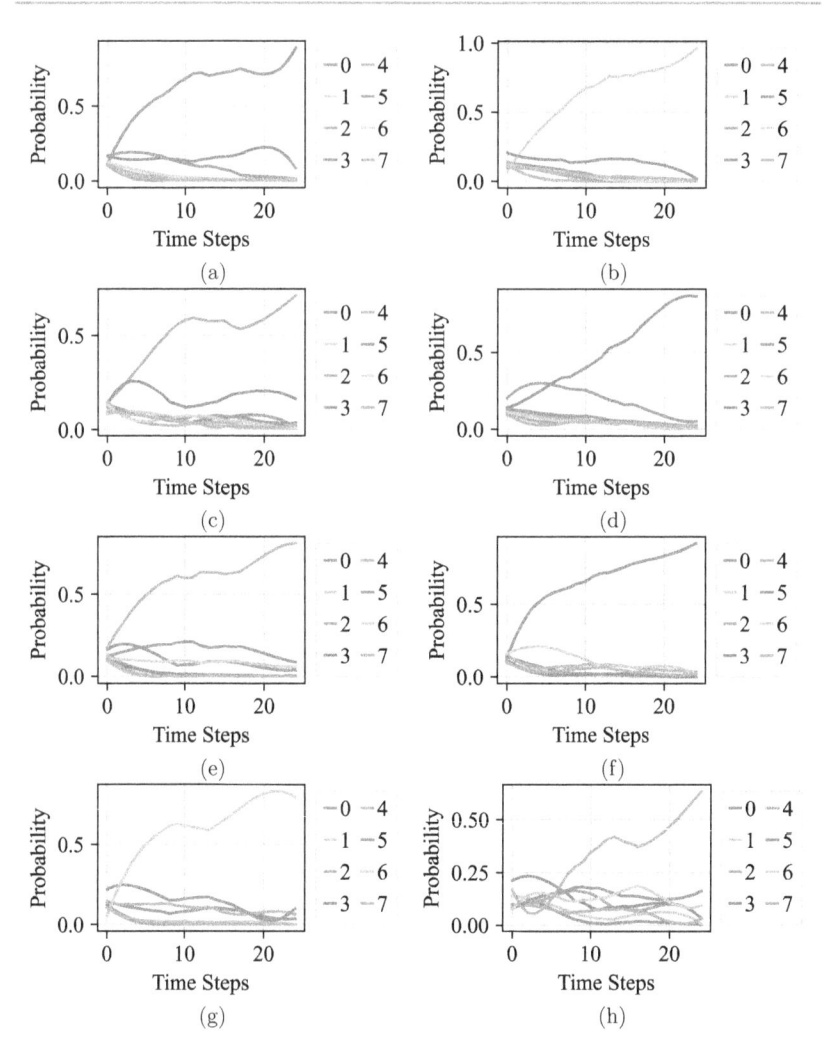

Fig. 3.24 Examples of the firing patterns of the SNN for each gesture class. The classes consist of: a) 0—Down Up, b) 1—Up Down, c) 2—Left-Right, d) 3—Rubbing, e) 4—Right-Left, f) 5—Diagonal Southwest to Northeast, g) 6—Diagonal Southeast to Northwest, h) 7—Clapping. It is evident that after a few time steps, the SNN starts to fire in accordance with the correct class, indicating its ability to accurately recognize and classify the gestures

3.8 Mimicking Fourier Transforms with Spiking Neural Networks

the model exhibits similar firing probabilities for each class. However, as the observation progresses, the firing probabilities converge towards a higher value for the correct class, indicating an increasing confidence in the prediction. This improvement in accuracy can be attributed to the integration of spikes over an extended duration, enabling the system to make more accurate class predictions.

As we aimed to develop a model that is both biologically plausible and computationally efficient, we chose to use the LIF model due to its simplicity. Similar to conventional deep neural networks or any other machine learning model, the performance of the SNN is highly dependent on its hyperparameters. In the case of SNNs, there are two types of hyperparameters to consider: those that apply to the overall SNN model and those that apply at the level of individual neurons. A grid search was performed to obtain the optimal parameters, which were used to train the network and consider both overfitting and underfitting. Table 3.12 and Table 3.13 show the optimal parameters used for the SNN model and neuron level, respectively. Additionally, Table 3.14 shows the optimal parameters used for Adam

Table 3.12 The hyperparameters employed during the training of the proposed SNN model

Parameters	Value
No. of epochs	10
Batch size	32
Learning rate (η)	0.011
Time steps during the testing	25
No. of dense layers	1
No. of dense layers units	64
No. of CNN layers Filters	1
No. of CNN Filters	16
Output synaptic delay	10 ms

Table 3.13 The parameters utilized for the LIF neuron in the proposed SNN

Parameters	Value
τ_{ref}	0.002
τ_{rc}	0.02
amplitude	0.001
min voltage	0

Table 3.14 The parameters utilized for Adam optimization and the weight initialization of the dense and convolutional layers

Parameter	Symbol	Value
Learning rate	α	0.0011
Decay rate for first moment estimate	β_1	0.9
Decay rate for second moment estimate	β_2	0.999
Epsilon	ϵ	$1e-8$
Dense layers (other than FFTS layer) weights	w_d	Drawn from normal distribution $N(0, 0.01)$
Convolutional layer weights	w_c	Chosen randomly from $\left[-\sqrt{\frac{6}{N_{in}+N_{out}}}, \sqrt{\frac{6}{N_{in}+N_{out}}}\right]$ where N_{in}, N_{out} are the number of input and output units

optimization and weight initialization of the dense and convolutional layers. The weights of the convolutional layer were initialized using the well-known Xavier initialization [111], which has been shown to be suitable for SNNs [112]. Using the parameters listed in Tables 3.12, 3.13, and 3.14, our proposed system achieves a gesture classification accuracy of 98.7%, which is similar to the accuracy levels achieved by state-of-the-art (earlier proposed in Sects. 3.4, 3.5, 3.6 and 3.7) SNNs and deepNets (as shown in Table 3.11).

The confusion matrix obtained by testing the model with the aforementioned dataset is depicted in Fig. 3.25(c). It can be observed that gestures 1 and 2 are classified correctly with an accuracy of 100% without any confusion. Gesture 3 shows only 1.10% confusion with gesture 5 and gesture 4 has 0.87% confusion with gesture 6 and 7 respectively. The highest misclassification rate of around 5.88% occurs for gesture 5, which is confused with gesture 7. Gesture 6 is misclassified as gesture 7 in 1.03% of cases, while gesture 7 is misclassified as gesture 6 in 0.88% of cases. Additionally, gesture 8 is confused with gesture 4 and 5, with a misclassification rate of 0.98% for each.

To demonstrate the performance improvement achieved by the proposed approach, which mimics range and Doppler FFTs, we conducted an ablation study. Specifically, we compared the performance of our approach with two other models: one without any FFTs (referred to as Model_1) and another that only mimicked the range FFT (referred to as Model_2). The confusion matrices depicted in Fig. 3.25 demonstrate the classification performance of the proposed system (Model_3) in

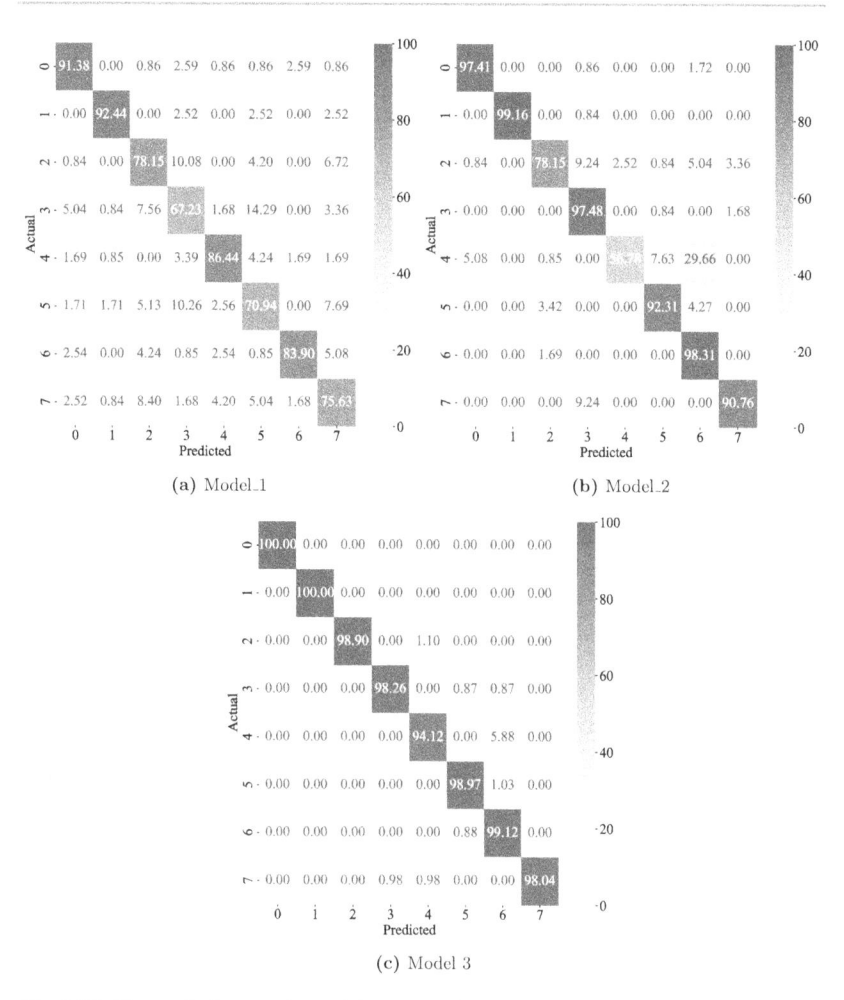

3.8 Mimicking Fourier Transforms with Spiking Neural Networks 103

(a) Model_1

(b) Model_2

(c) Model 3

Fig. 3.25 The confusion matrices of the models listed in Table 3.15 using the test dataset. Model_3 (our proposed model) shows good performance compared to the other models, due to the first layers mimicking the FFTs. The values in the confusion matrices are shown in percentages, and the axes correspond to the 8 classes of gestures: 0—down up, 1—up down, 2—left-right, 3—rubbing, 4—right-left, 5—diagonal southwest to northeast, 6—diagonal southeast to northwest, 7—clapping

comparison to Model_1 (Fig. 3.25(a)) and Model_2 (Fig. 3.25(b)). Notably, the proposed model exhibits significantly lower levels of confusion among different classes. This improvement can be attributed to the incorporation of FFT mimicking in the initial layers of the network, which enables the system to effectively capture and utilize spatiotemporal information for enhanced classification accuracy.

To further comprehend on the evaluation of our proposed system, we utilized the t-Distributed Stochastic Neighbour Embedding algorithm (t-SNE) to visualize the feature space and assess the discriminative capabilities of the network for each class in the dataset. In this analysis, we fed the output of the last layer (prior to the classification layer) into the t-SNE algorithm, along with the corresponding class labels. By leveraging t-SNE, we aimed to gain insights into the clustering and separability of the features generated by our network, providing valuable information about the discriminative power of the proposed system. We conducted a systematic investigation by gradually increasing the number of layer neurons from 2 to 64 in powers of 2. For each configuration, we calculated the t-SNE visualization, as shown in Fig. 3.26. The results demonstrate a clear trend: as the number of neurons increases, the separability of the features improves. Notably, our proposed SNN exhibits the ability to learn distinct and separable features, as evidenced by the tightly clustered groups observed in the t-SNE plot. Specifically, the SNN achieved accurate classification of the 8 gesture types using a lower dimension of 32 neurons. This finding highlights the efficiency and effectiveness of our SNN architecture in extracting meaningful representations for gesture classification tasks.

Additionally, Fig. 3.27(c) illustrates the discriminating features that were learned by each layer. It is noticeable that as we move along the layers, the features become increasingly discriminating and form close clusters. At the final layer, the features are highly discriminative and thus are effectively classified. Table 3.15 clearly demonstrates the significance of incorporating range and Doppler FFTs in the SNN model. It presents the accuracy performance of the model with and without FFTs during both the training and testing phases. It is evident that the model's accuracy is not satisfactory, reaching only 81.03% without the FFTs. However, by introducing the range FFT layer, the performance improves by 7.87%. The proposed method, which includes the Doppler FFT layer, further enhances the accuracy to 98.7%. Additionally, the t-SNE plots of the CNN model (Fig. 3.27(a)) and range model (Fig. 3.27(b)) demonstrate that the features generated by these models are not well-discriminating, leading to lower performance compared to the proposed model.

In order to enhance power efficiency even more, we examined how post-training quantization affects accuracy performance, as demonstrated in Table 3.16. Quantization has a two-fold impact on power efficiency: it decreases both memory footprint costs and computational costs. Moreover, the utilization of quantized data with

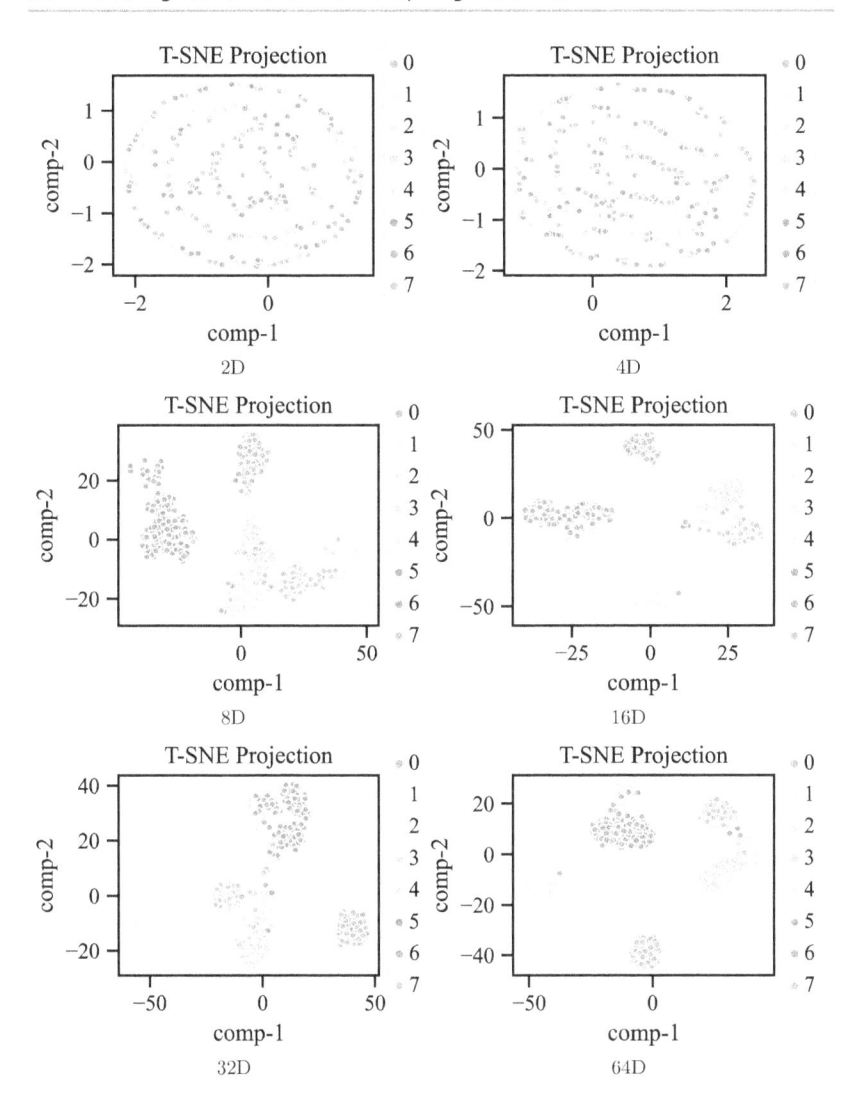

3.8 Mimicking Fourier Transforms with Spiking Neural Networks

Fig. 3.26 The t-SNE plots that depict the dimensionality reduced embedded feature clusters in varying dimensions, ranging from 2 to 64. The subfigure (e) highlights that even in lower dimensions, the SNN model can learn separable and discriminating features

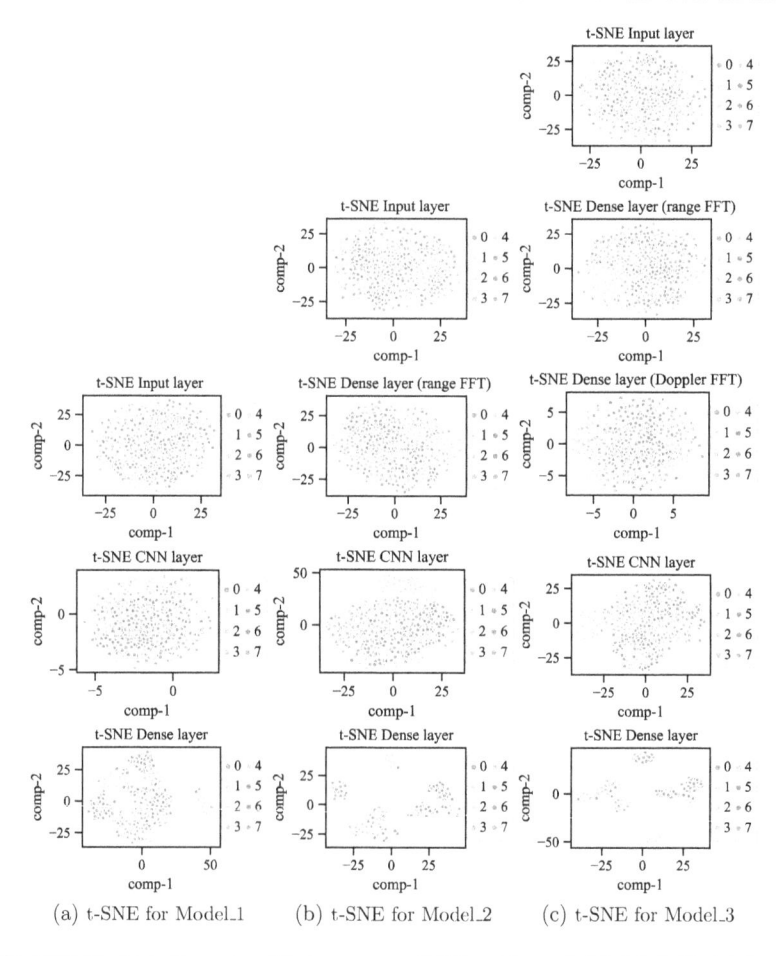

(a) t-SNE for Model_1   (b) t-SNE for Model_2   (c) t-SNE for Model_3

Fig. 3.27 The high-dimensional feature spaces of different layers in Model_1, Model_2, and Model_3. In a), the t-SNE visualization is shown for the input, CNN, and dense layers of Model_1. b) depicts the t-SNE for the range model (Model_2), while c) illustrates the t-SNE for the layers of the proposed method (Model_3). The legend in each sub-figure corresponds to the classes as follows: 0—down up, 1—up down, 2—left-right, 3—rubbing, 4—right-left, 5—diagonal southwest to northeast, 6—diagonal southeast to northwest, 7—clapping. It can be observed that as we move deeper into the network, the features become more discriminative, leading to improved separability among different gesture classes

Table 3.15 A comparative analysis of classifications based on various layers, utilizing identical parameters as depicted in Tables 3.12, 3.13, and 3.14

Model	Layers	Accuracy
Model_1	Input	81.03%
	Convolution (16 filters, 3 kernel size)	
	Dense (64)	
	Dense (8)	
Model_2	Input	88.90%
	2 × Dense (64)	
	Convolution (16 filters, 3 kernel size)	
	Dense (64)	
	Dense (8)	
Model_3	Input	98.7%
	2 × Dense (64)	
	4 × Dense (64)	
	Convolution (16 filters, 3 kernel size)	
	Dense (64)	
	Dense (8)	

reduced bit rate offers several advantages, including enhanced energy efficiency and reduced memory bandwidth requirements. This is due to the reduced need for on-chip and off-chip data movement when compared to higher bit rates. The effect of post-training quantization on the three models listed in Table 3.15 is demonstrated in Table 3.16. In the case of 4-bit quantization, 1 bit is allocated for the sign, 2 bits for the integer part, and 1 bit for the fractional part. Likewise, for 8-bit and 16-bit quantization, 1 bit is assigned to the sign, 3 bits are dedicated to the integer part, and the remaining bits are utilized for the fractional part. As anticipated, the accuracy declines as the quantization level increases, particularly for our proposed model (Model_3), with a greater percentage decrease. The observed reduction in accuracy can be attributed to the increased number of neurons in the proposed model, which amplifies the impact of quantization. However, we acknowledge that higher accuracy levels could potentially be achieved through the use of quantization-aware training techniques. It is important to note that the current version of the nengoDL framework does not support such techniques.

The accuracy of the proposed model demonstrates improvement as the number of bits used for quantization increases. Specifically, with 8-bit and 16-bit

Table 3.16 The impact of post-training quantization on accuracy

Model	Quantization	Accuracy	f1-score_micro	f1-score_macro
Model_1	4-bit	78.10%	0.781	0.7821
	8-bit	81.38%	0.8138	0.8143
	16-bit	80.74%	0.8074	0.8085
Model_2	4-bit	78.94%	0.7894	0.7889
	8-bit	88.57%	0.8857	0.822
	16-bit	88.57%	0.8857	0.826
Model_3	4-bit	77.20%	0.772	0.7619
	8-bit	92.68%	0.9268	0.9222
	16-bit	96.46%	0.9646	0.9631

quantization, the model achieves accuracies of 92.68% and 96.46%, respectively. The precision and recall metrics, as indicated by the f1-scores, also exhibit excellent performance. In the case of "f1-scores_micro," the average is calculated by considering the total number of true positives, false positives, and false negatives. Similarly, "f1-scores_macro" computes the metric for each class and then calculates their unweighted arithmetic mean.

For energy efficiency, we examined the energy consumption per classification of our proposed system. Since conducting actual hardware-based energy measurements falls outside the scope of this research, we relied on the hardware metrics provided by the μBrain chip, as documented in [108], to estimate the energy consumption in our study. If σ represents the maximum number of spikes, $\eta = 2.1$ pJ denotes the energy per spike, and $\Lambda = 73\,\mu\text{W}$ represents the static leakage power, the energy consumption per classification can be calculated using the μBrain hardware metrics as follows:

$$\omega = \sigma \times \eta + \delta T \times \Lambda, \quad (3.40)$$

where ω represents the energy consumption per classification. Assuming an inference time of $\delta T = 28$ ms, the estimated energy consumption per classification for our proposed system is approximately $\omega = 2.1\,\mu\text{J}$.

For a comparison of energy efficiency in SNN hardware, readers can refer to [113], which compares SNN hardware to other deep learning hardware based on their energy efficiency. The SNN hardware's performance was evaluated on a keyword spotting application using some commercially available energy-efficient accelerators, as depicted in Fig. 3.28, with a dynamic energy cost per inference. The dynamic energy cost per inference, as defined in [113], represents the energy

Dynamic Energy Cost Per Inference (batchsize=1)

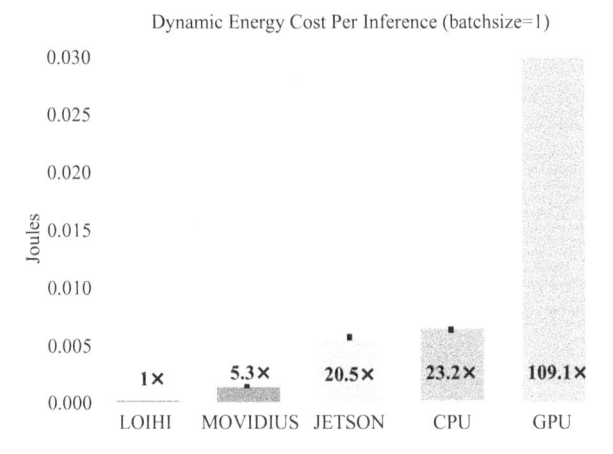

[Bar chart showing Joules on y-axis (0.000 to 0.030) with bars for:
- LOIHI: 1×
- MOVIDIUS: 5.3×
- JETSON: 20.5×
- CPU: 23.2×
- GPU: 109.1×]

Fig. 3.28 Comparison of energy efficiency in SNN hardware with other deep learning hardware for keyword spotting. The dynamic energy cost per inference was evaluated using commercially available accelerators. The experiment showed up to 10x improvement in power efficiency for SNN hardware [113]

consumed by the hardware during a single inference, excluding the energy consumed during idle periods. In [113], inference refers to the process of feeding an input vector through a two-hidden-layer feed-forward artificial neural network (ANN) to predict a probability distribution for alphabetical characters. The experimental results showcased a remarkable improvement in power efficiency, with up to a 10× reduction in energy consumption. The proposed prototype solution based on SNNs demonstrates remarkable accuracy in real-time hand gesture recognition, achieving performance on par with state-of-the-art deepNets and other SNN solutions. In addition, the utilization of SNNs in the proposed system ensures low power consumption, making it an ideal choice for applications where energy efficiency is required.

3.8.4 Conclusion

The proposed gesture sensing system introduces a novel approach by integrating an SNN with a 60-GHz radar system. In contrast to conventional methods that rely on image-based or point cloud input data, our system directly utilizes the raw ADC data as input for the SNN. This unique approach offers several advantages, including the implicit mimicking of Fourier transform processing (slow-time, fast-time). As a result, the system eliminates the need for additional FFT accelerators and tailors the FFT pre-processing specifically to the task of gesture sensing. This not only reduces computational overhead but also enhances the energy efficiency of the system. Compared to our proposed system in Sect. 3.6 and the state-of-the-art methods, the proposed SNN achieves a similar level of accuracy in recognizing 8 different hand gestures making the system a promising solution for embedded applications that require low latency and low power consumption.

3.9 Spiking Neural Networks-based Gesture Sensing on FPGA

In this section, we present our FPGA-based demonstration of SNN gesture sensing. The demonstration is illustrated in Fig. 3.29, where an FPGA-controlled lamp showcases the functionality of the SNN. The lamp can perform four distinct functions corresponding to four different gestures: turning on, turning off, changing color, and adjusting brightness.

Fig. 3.29 FPGA based gesture sensing demonstration using SNN to control a lamp

The radar system initially collects raw data, which is then transmitted to a PC via USB for signal processing tasks. These tasks involve generating a comprehensive map of range over time, velocity over time, and angle over time. The resulting map is subsequently converted into spike representations and sent to the FPGA through an ethernet connection.

Within FPGA, the SNN model is executed to process the received spikes and send the output spikes are relayed back to the PC via ethernet. The PC then processes the

3.9 Spiking Neural Networks-based Gesture Sensing on FPGA

output spikes and makes predictions based on the recognized gestures. Subsequently, the PC sends appropriate commands to the hue device associated with the lamp, triggering the desired gesture. The hue device, using the Zigbee [114] wireless protocol, conveys the corresponding signals to the lamp through a hue bridge, thus achieving wireless control over the lamp's behavior.

Figure 3.30 illustrates the deployment pipeline specifically designed for the SNN FPGA board. This pipeline is organized into four stages, namely model building, model optimization, model transformation, and the final deployment onto the hardware setup. In the subsequent subsection, we will take a closer look at each of these stages, offering a comprehensive explanation of their importance and the specific tasks associated with them.

Building a model in Nengo → Model Optimization → Model transformation to lower level C code → Deployment on the FPGA and host

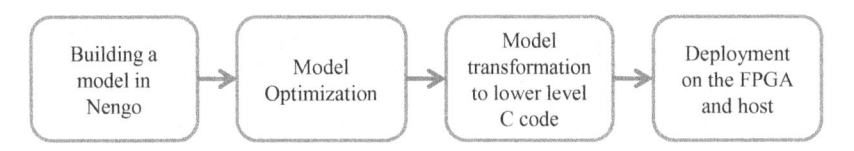

Fig. 3.30 The deployment pipeline created for the SNN FPGA board

3.9.1 Building a Model in Nengo

In our initial approach, we utilized the model introduced in Sect. 3.4. While the Nengo version of SNN demonstrates commendable performance in terms of classification accuracy, the size of the model surpasses the capacity of the FPGA emulation. This increased memory footprint is a consequence of an unoptimized structure encompassing weights, bias matrices, neuron states, and spike events. Consequently, adjustments need to be made to the input layer and hidden layers. However, simply reducing the size of the input leads to a significant decrease in accuracy. To address this challenge, we have incorporated the angle of arrival as an additional feature to the input. By incorporating this feature, the input format takes the shape of a 3D array comprising parameters such as range, radar Doppler, and angle over time, with dimensions of $14 \times 14 \times 3$. This modification reduces the input size from 2048 to 588 for the employed algorithms.

Subsequently, leveraging the reduced input size, we are able to incorporate the next layer in the network, which comprises a convolutional layer consisting of LIF neurons. This layer serves as an encoder, utilizing a kernel size of 2 and a filter size of 3. The purpose of this layer is to effectively reduce the data size to 147 which can then be interpreted as spike inputs by subsequent layers within the network.

Following the encoding process, the output of 147 binary numbers is connected to a dense layer containing 32 LIF neurons. This layer is succeeded by another dense layer with 4 neurons, which serves as the output layer. Additionally, a Nengo synapse is incorporated, representing a linear filter. The optimized architecture of the SNN is visually depicted in Fig. 3.31. Remarkably, this network exhibits a comparable level of performance in terms of accuracy (98.5%), all while maintaining an impressively low memory footprint.

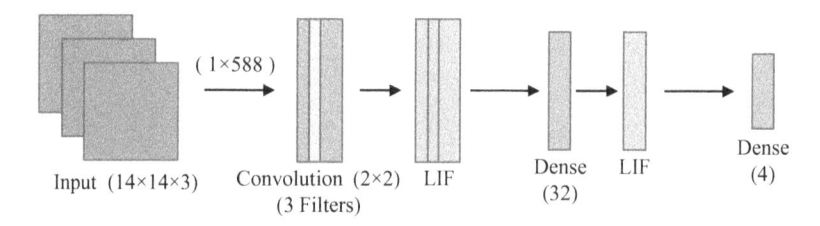

Fig. 3.31 The optimized architecture for FPGA: Begins with an input size of (14 × 14 × 3), followed by a convolutional layer using a (2 × 2) kernel and 3 filters, succeeded by a dense layer featuring 32 neurons, and concluding with a final dense layer having 4 neurons. LIF activation is applied to all layers except the final one

3.9.2 Model Transformation to Lower Level Code

This stage involves transforming the optimized model into lower-level code suitable for deployment on the FPGA. This transformation is necessary due to the lack of support for Xilinx FPGAs in the Nengo simulator. Consequently, the Nengo functions utilized in the model would be incompatible with the platform. To accomplish this transformation, a Python script was initially developed, designed to minimize dependencies on high-level abstract libraries. This script effectively replicates the behavior of the Nengo model using concise code that employs low-level, easily traceable operations. Throughout the development process, the script's behavior was validated by comparing the layer-by-layer output, specifically the spike activities, with those of the Nengo reference model. Once the Python code was validated, it was rewritten in C and adapted for FPGA implementation.

To streamline the process, a Python script was created to automate the transformation of a Nengo model into C code. This script autonomously handles the conversion, eliminating the need for manual intervention. Following that, the FPGA implementation involves the incorporation of any remaining hidden layers along with their

3.9 Spiking Neural Networks-based Gesture Sensing on FPGA

corresponding LIF neurons. Additionally, certain pre-processing steps are applied on the board, configured to apply weighting to incoming spikes prior to forwarding them to the LIF neurons. Finally, the output layer is seamlessly integrated into the Linux host component of the application pipeline.

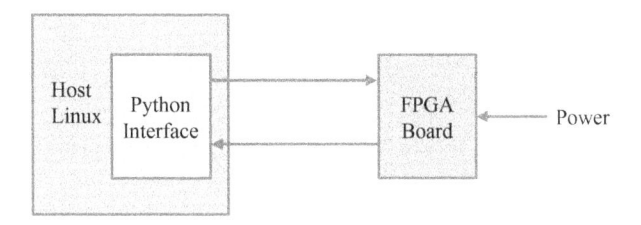

Fig. 3.32 Schematic diagram of the FPGA implementation

3.9.3 Implementation on the FPGA

With the model in C code format, the implementation or direct deployment on the FPGA is started. The schematic diagram of the deployment of the FPGA is shown in Fig. 3.32. Where a Linux host is connected to an FPGA board with a python interface for communication/transfer of the data from Host to FPGA and vice versa. The input and the convolutional layer are implemented on the host side. The spike activity is then sent to the FPGA development board. The hidden layer with 32 neurons is programmed on the FPGA and then the output spike activity of the layer is sent back to the host. Where the last output layer of 4 neurons with SoftMax activation function performs the classification task.

3.9.4 Conclusion

The deployment process highlighted the immaturity of state-of-the-art tools for SNNs when it comes to implementing them on hardware prototypes. This conclusion was drawn based on the extensive amount of time required to transform the Nengo model into a format that could be efficiently executed on both the host and the SNN FPGA platform. Our Xilinx FPGAs implementation is able to achieve a comparable level of performance in terms of accuracy 98.5%.

Radar-based Air-writing for Embedded Devices

4.1 Introduction

Air-writing (AW) is an advanced type of gesture sensing (GS) that enables users to write in a 3D space by moving their hand or a marker. It involves writing characters, numerals, symbols, and words without any physical medium. Unlike hand gestures, which can be deliberately made to facilitate recognition, air-writing involves varying directions, speeds, and styles, making it more challenging to recognize. AW has been implemented using different types of sensors, and the systems can be categorized based on the sensor used. Similar to GS, AW can be classified into two main categories, which are vision-based and non-vision-based sensors. Further categorization of specialized techniques can be made based on the type of sensor interaction, as illustrated in Fig. 4.1.

AW techniques that rely on vision-based sensors, including cameras, stereo-vision sensors, and depth sensors, are often powered by deep learning and machine learning methods [115–117]. Although these techniques are prevalent in comparison to other sensors, they are susceptible to illumination variations and privacy concerns. The data processing needed for these methods is complex, making them unsuitable for portable or edge devices. The first category of non-vision-based techniques involves wearable devices such as gloves, smartwatches, and smartphones that use inertial sensors [118–120]. While the aforementioned sensors offer the advantages of being low-cost and compact, their wearable nature may prove cumbersome to users. As a result, non-contact or contactless AW techniques were developed. These techniques involve projecting media onto a target and extracting information from the reflected media. In recent years, researchers have explored ultrasonic-based waves for AW which offer the advantage of tracking with millimeter-level accuracy

```
                        Air-writing
            ┌───────────────┴───────────────┐
       Vision Based                   Non-vision Based
    ┌───────┼────────┐              ┌──────┴──────┐
 Cameras Stereo-vision Depth Sensors Contact Less  Contact Based
                                    ┌──────┴──────┐      │
                                  Acoustic    Radio Frequency  Wearable
                                   Based         Based         Sensors
                                              ┌───┴───┐
                                         Non-radar Based  Radar Based
```

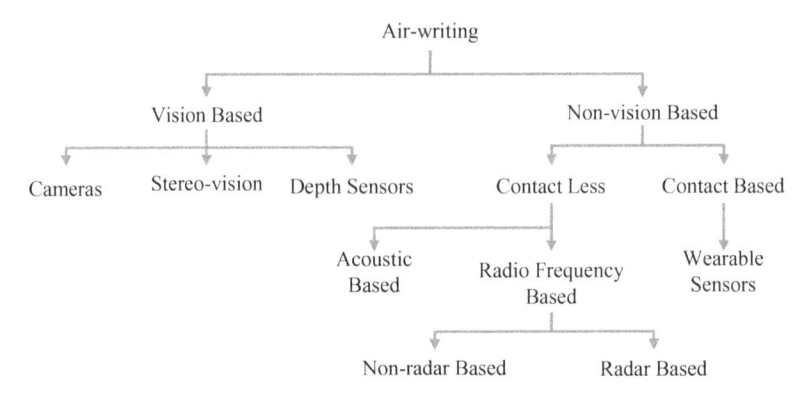

Figure 4.1 Categorization of air-writing techniques based on sensor used, with vision-based and non-vision-based sensors as the two main categories

at short ranges [121–123]. However, the accuracy decreases as the distance between the target and the sensors increases.

In recent times, the use of radio frequency (RF) based sensors for AW applications has garnered significant attention from the research community. The RF-based techniques can be categorized based on the RF band used. The non-radar based AW technologies are primarily based on Wireless Fidelity (WiFi) [124–126]. However, the setup of these systems requires several transmitters and receivers with large antenna spacing, making them impractical for real-life applications. Moreover, the recognition performance of these techniques suffers from high interference within the band used. Radar advances the aforementioned sensors because of their compact size, penetrating capabilities and lower memory footprint making them suitable for portable IoT devices.

Recently, its potential has been shown in AW applications with great performance capabilities [127], [128]. However, mostly they are driven by the learning capabilities of deep learning [127], [129], [130], require multiple radars [127] that require huge computation making them energy inefficient. The method [128] requires a physical pad or board to write, therefore, disqualifying it from the category of AW.

4.1.1 State-of-the-Art

The state-of-the-art air-writing systems [127] use a network of millimeter-wave FMCW radars. A stroke represents a writing trajectory performed through hand

4.1 Introduction

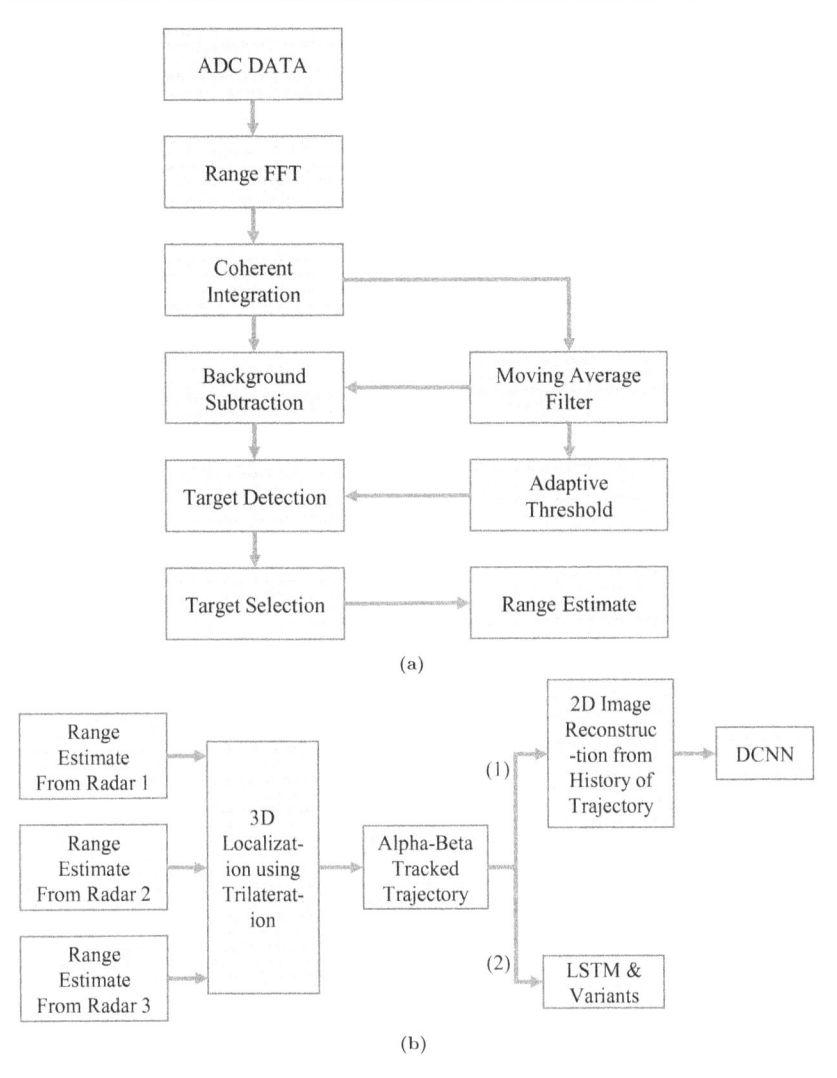

Figure 4.2 Overview of an air-writing system using millimeter-wave FMCW radars. a) Steps for range estimation: FFT spectral analysis, coherent integration, and MTI filtering. b) Trilateration for precise hand marker estimation in 3D coordinates, followed by motion tracking with a smoothening filter. Hand motion is transformed to a 2D image for DCNN gesture classification or tracked marker coordinates are input to LSTM for classification [127]

motions in the air. Accurate localization of hand markers is vital for estimating gesture trajectories. Trilateration techniques ensure precise estimation of hand markers in three-dimensional coordinates, and a smoothing filter tracks the hand trajectory. In the recognition stage, a textual representation is generated from sensor data capturing hand motion. This representation is input into advanced deep networks like LSTM and its variants, alongside DCNN. The conventional solution is depicted in Fig. 4.2 and summarized below.

4.1.1.1 Range Estimates

To obtain estimates of the time delay in FMCW radar, FFT spectral analysis is performed along each chirp, referred to as fast time. The resulting FFTs for all chirps in a frame are then combined using pulse integration. This coherent integration improves the SNR and enables efficient target detection. Coherent integration at frame k, $R_{ci}(k)$, is based on the mean of the range FFT data $\{R_i^n(k)\}_{i=1}^{N_C}$ over N_C chirps and is given as:

$$R_{ci}^n(k) = \frac{1}{N_C} \cdot \sum_{i=1}^{N_C} R_i^n(k). \tag{4.1}$$

To filter out stationary objects and reveal only moving targets, a simple one-dimensional MTI filter is applied that involves the subtraction of the moving average of the range spectrum. Only moving targets are exclusively captured in the residual range spectrum. After applying the MTI filter, target detection and selection are accomplished by applying a threshold to the scaled mean value of the residual spectrum, and subsequently selecting the nearest target. Upon obtaining range information from all three radars, a trilateration algorithm is employed to calculate the global 3D coordinates relative to a reference point.

4.1.1.2 Target Localization with Trilateration

We utilized the trilateration method, as outlined in the reference [131], to calculate the global coordinates of an object in space. Trilateration involves leveraging distance measurements from a minimum of three pre-defined reference points with known coordinates. To demonstrate the trilateration technique, let us consider three reference points, denoted as $R_1(i_1, j_1, k_1)$, $R_2(i_2, j_2, k_2)$, and $R_3(i_3, j_3, k_3)$. The location or coordinates of a target point $T(i, j, k)$ can be represented by the distances d_1, d_2, and d_3 from R_1, R_2, and R_3, respectively, by solving the following system of quadratic equations:

4.1 Introduction

$$\begin{cases} (i - i_1)^2 + (j - j_1)^2 + (k - k_1)^2 = d_1^2 \\ (i - i_2)^2 + (j - j_2)^2 + (k - k_2)^2 = d_2^2 \\ (i - i_3)^2 + (j - j_3)^2 + (k - z_3)^2 = d_3^2. \end{cases} \quad (4.2)$$

In matrix form, this can be expressed as:

$$\begin{bmatrix} 1 & -2i_1 & -2j_1 & -2k_1 \\ 1 & -2i_2 & -2j_2 & -2k_2 \\ 1 & -2i_3 & -2j_3 & -2k_3 \end{bmatrix} \begin{bmatrix} i^2 + j^2 + k^2 \\ i \\ j \\ k \end{bmatrix} = \begin{bmatrix} d_1^2 - i_1^2 - j_1^2 - k_1^2 \\ d_2^2 - i_2^2 - j_2^2 - k_2^2 \\ d_3^2 - i_3^2 - j_3^2 - k_3^2 \end{bmatrix}. \quad (4.3)$$

Above Equation (4.3) can be simply written in the form:

$$\mathbf{Ax} = \mathbf{b}, \quad (4.4)$$

Subject to the constraint $x \in C$, where,
$C = \left\{ (i_0, i_1, i_2, i_3)^T \in \mathbb{R}^2 / i_0 = i_1^2 + i_2^2 + i_3^2 \right\}.$
The problem mentioned above can be solved efficiently and has been briefly described in reference [131]. We have utilized the solution to generate baseline reference trajectories [127], which will be compared to our proposed solution.

4.1.1.3 Trajectory Smoothening Filters

After obtaining the global coordinates, the trajectory data undergoes a smoothing process using an $\alpha\beta$ filter to mitigate any disturbances caused by noise. The prediction process is formulated as follows:

$$p(n) = \bar{p}(n-1) + T_f \cdot v(n-1), \quad (4.5)$$

$$v(n) = \bar{v}(n-1), \quad (4.6)$$

In the above equations, T_f represents the frame time or measurement update interval. $p(n)$ and $\bar{p}(n)$ indicate the predicted and smoothed target position at time nT_f, while $v(n)$ and $\bar{v}(n)$ represent the predicted and smoothed target velocities at time $t_n = nT_f$, respectively. The update process is defined as follows:

$$\bar{p}(n) = p(n) + \alpha(\hat{p}(n) - p(n)), \quad (4.7)$$

$$\bar{v}(n) = v(n) + \frac{\beta}{T_f}(\hat{p}(n) - p(n)), \quad (4.8)$$

Here, $\hat{p}(n)$ denotes the measured position of the target at the trajectory instant nT_f.

4.1.1.4 Trajectory Reconstruction and Classification

After smoothing the trajectory, two alternative approaches can be taken. The first generates a full 2D trajectory image by detecting the end of hand motion and inputting the feature image into a DCNN (Deep Convolutional Neural Network). The second approach feeds the smoothed trajectories into an LSTM (Long Short-Term Memory) network for character recognition.

4.1.2 Limitations of State-of-the-Art

The state-of-the-art systems rely on the utilization of multiple radars, typically three or more, to perform trilateration and estimate the global coordinates of a moving trajectory. However, practical implementations often encounter several limitations. These include issues with the trilateration algorithm due to occlusion caused by fingers obstructing one or more radars, as well as missed detections leading to incomplete or unreliable trajectory coordinates. Furthermore, the practicality of optimizing the intersecting field-of-view (FoV) by arranging a network of radars on a computer screen or AR-VR device is often hindered by the restricted FoV of each radar unit and the difficulties associated with their physical placement. Moreover, the conventional approach of using 2D trajectory images as feature inputs for classification using DCNNs fails to fully exploit the capabilities of these powerful models, which have the potential to implicitly learn relevant features. Moreover, the conventional methods rely on deepNets that demand substantial computational resources, rendering them energy-inefficient and unsuitable for embedded devices.

4.1.3 Proposed Solutions

In this chapter, we propose various systems using a sparse network of radars, specifically fewer than three. These systems exploit solely the range data provided by each radar and employ compact deepNets to simultaneously extract features, akin to trilateration, and model the local radar trajectories based on information from one or two radars. As a result, these systems offer a solution with low memory requirements, rapid inference time, and enhanced accuracy. Additionally, to tackle the energy efficiency issues of deepNets, we proposed SNNs-based AW systems that are highly energy efficient. The energy efficiency of SNNs is attributed to their transmission process, which occurs only when a neuron's membrane potential reaches a predetermined threshold. Otherwise, the neuron remains in an idle state. Consequently, the communication within the network is greatly sparse, contributing to its energy-

4.1 Introduction

efficient nature. Moreover, the transmission in this context is represented by 1-bit activity, and the conventional Multiply-Accumulate (MAC) arrays are substituted with adders that solely accumulate the incoming spike. As a result, these computational operations are exceedingly minimal, further reducing the overall computational load. Given their low power consumption, fast inference capabilities, and event-driven processing, SNNs are highly suitable for implementing deepNets and machine learning tasks, particularly in scenarios where energy efficiency is a critical consideration. Despite the energy efficiency of the SNNs [96–99], the use of the SNN is limited because the training of the SNNs is a challenge because of the non-differentiable activation functions not allowing backpropagation. Among various learning mechanisms, the conventional backpropagation technique is typically modified for deep SNNs. This involves initially training the network using differential approximated activation functions and subsequently replacing these functions with spiking activation functions during the testing phase, thereby transforming the network into an SNN. In our work, we adopt the same approach.

The main contribution of this chapter are as follows:

- In Sect. 4.4.2 we propose an air-writing system with the following contributions:
  - We propose an air-writing system that utilizes a sparse network of radars, specifically employing only one or two radars.
  - A novel architecture 1D DCNN-LSTM-1D transpose DCNN is proposed to enable the simultaneous recognition of the drawn character and reconstruction of the 2D trajectory image.
  - When using hand-crafted feature images generated through trilateration or extended Kalman filter, additional data augmentation or preprocessing steps are typically necessary to achieve invariance to the reference of the imaginary board. In contrast, the proposed system has the advantage of being inherently immune to such transformations due to the intrinsic learning of feature images by the deepNet.
  - By utilizing real 60-GHz sensor data, we showed air-writing recognition system that exhibits remarkable accuracy. This achievement paves the way for effortless practical implementation of similar systems on inexpensive hardware, making them more accessible and widely deployable.
- In Sect. 4.5.2 we propose an air-writing system with the following contributions:
  - We propose an air-writing system that utilizes only one or two radars, allowing the input to the classifier to be based on the local location of the hand or marker, rather than the global location.

- We propose a novel 1D TCN (Temporal Convolutional Network) architecture that simultaneously performs feature extraction and recognition of locally drawn radar trajectories.
- The proposed system is capable of handling both individual characters and continuous writing.
- The practical applicability of the proposed air-writing recognition based on the local trajectory is demonstrated using data from Infineon's 60-GHz FMCW radar sensor showing the potential for deploying such solutions in space or cost-constrained systems.
- In Sect. 4.6.2 we propose an air-writing system with the following contributions:
  - To our knowledge, this is the first instance of utilizing SNN in the context of air-writing.
  - We present a novel SNN architecture for the classification of drawn trajectories. The proposed architecture utilizes a network of three radars in conjunction with trilateration techniques.
- In Sect. 4.7.2 we propose an air-writing system with the following contributions:
  - We propose a novel air-writing approach using a single radar and Spiking Legendre Memory Unit (SLMU) that makes the solution compact and energy efficient on an edge device.
  - A novel SNN architecture is proposed for the classification of drawn trajectories with only one or two radars.
  - We propose the use of a genetic algorithm to get the optimal hyperparameters of the SNN.

The rest of the chapter is organized as follows: Sect. 4.2 presents the radar hardware used in this chapter, and Sect. 4.3 presents the dataset collection and experimental setup. Sect. 4.1.1 presents the conventional air-writing approach. The Sects. 4.4.2 4.5.2 4.6.2 and 4.7.2 presents our proposed air-writings systems.

4.2 Hardware

We have utilized Infineon's BGT60TR24B 60-GHz FMCW radar chip to implement our system, as illustrated in Fig. 4.3(a). Fig. 4.3(b) depicts a typical block diagram of FMCW radar. The operating frequency of the radar is between 57 GHz to 64 GHz and the chirp duration can be adjusted. The linear frequency sweep is controlled externally by a phase-locked loop (PLL). The PLL uses the frequency divider output pin and the tuning voltage pin to generate the frequency chirps. The

4.2 Hardware

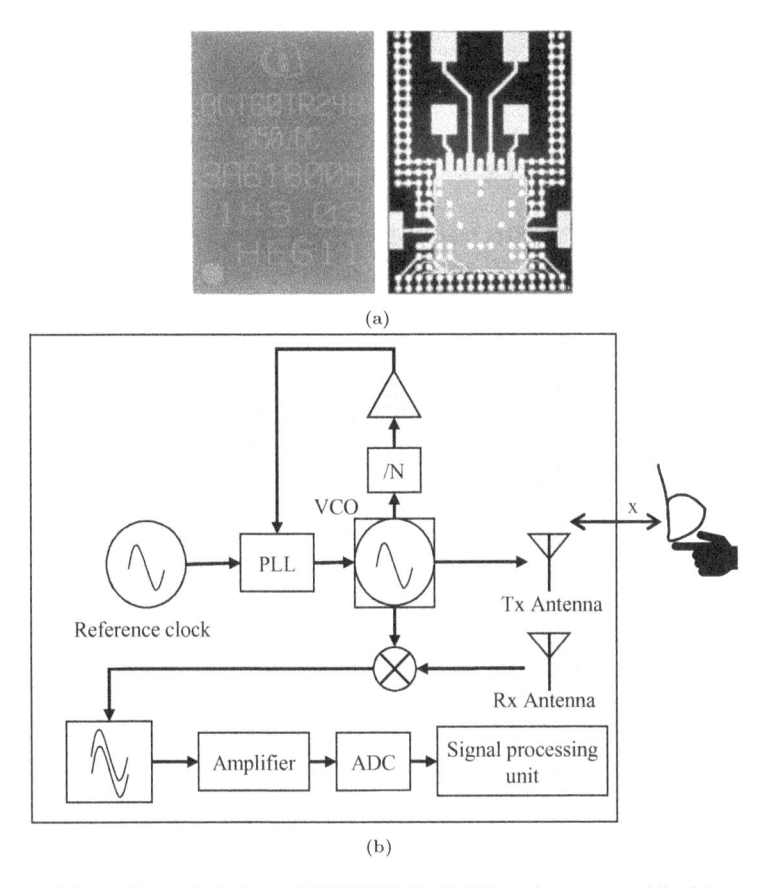

Figure 4.3 (a) Shows the Infineon BGT60TR24B 60-GHz radar sensor, while (b) presents a simplified block diagram of the chipset

tune voltage output is adjusted according to the divider output with an 80 MHz reference oscillator, and a voltage-controlled oscillator (VCO) generates the chirps. The VCO tunes the voltage from 1 to 4.5 V, resulting in highly linear frequency chirps between 57 GHz and 64 GHz. The reflected signal from the target is mixed with a copy of the transmitted signal to obtain an intermediate signal. The signal is then passed through low-pass filters before being sampled by the analog-to-digital converter (ADC), and only the in-phase (I) channel is sampled. Additionally, each receiver path is fed through a low-noise voltage-gain amplifier and a high-pass filter. FMCW radars require significantly reduced sampling rates for the ADC and subsequent signal processing, making them compact and cost-effective.

4.2.1 System Parameters

We used the parameters depicted in the able 4.1 below to configure the radar for our experiments.

Table 4.1 The radar system parameters employed within this study

Parameters with symbol	Value
Range resolution of single radar	2.5 cm
Elevation θ_{elev} per radar	70°
Azimuth θ_{azim} per radar	70°
Ramp start frequency, f_{min}	57 GHz
Ramp stop frequency, f_{max}	63 GHz
Bandwidth, B	6 GHz
Chirp time, T_c	171.2 μ sec
Sampling frequency, fs	0.747 664 MHz
Number of samples per chirp, Ns	128

4.3 Dataset and Experimental Setup

To demonstrate our proposed system, we employed the dataset from [127], containing 10 alphabets (from A to J) and 5 numerals (from 1 to 5). Data collection occurred in two setups: one with radars forming a horizontal virtual board (Fig. 4.4(b)) and another with a vertical board (Fig. 4.4(a)). ADC data was recorded on a PC connected to three radars via USB. Recording encompassed 20 consecutive frames after detecting the hand marker or target by all radars. Characters drawn for fewer than 20 consecutive frames were zero-filled. Examples of 2D reconstructed trajectories for each class from the dataset are presented in Fig. 4.5.

4.3 Dataset and Experimental Setup

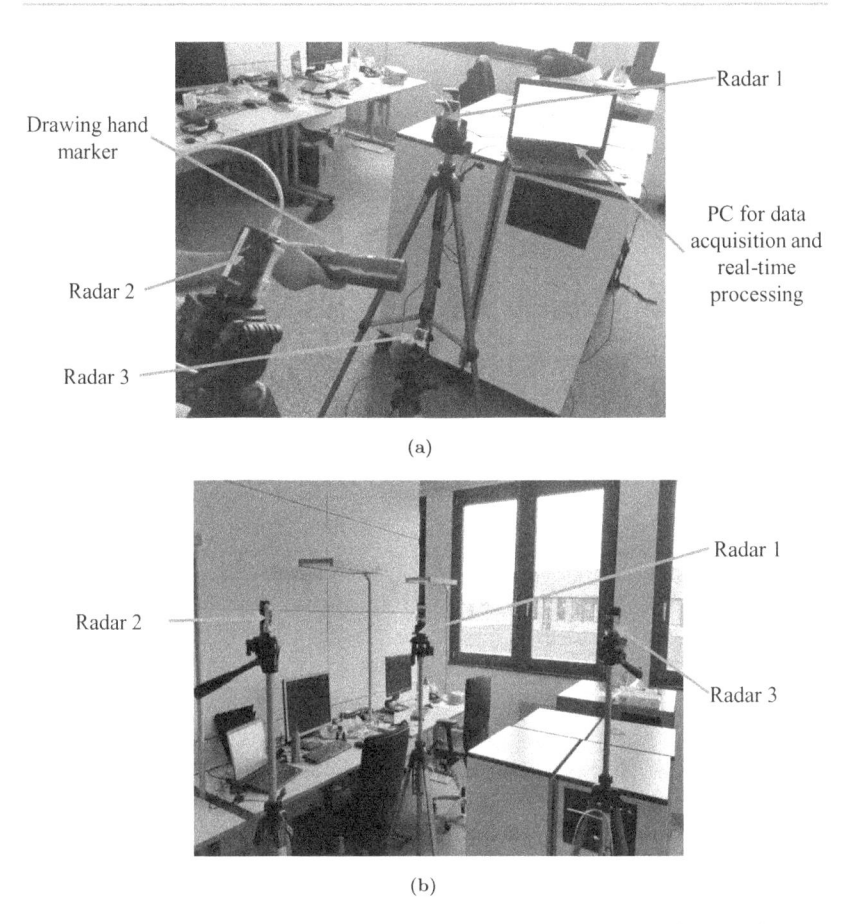

Figure 4.4 Figure depicting the two data collection scenarios: (a) the radars positioned to create a vertical virtual board, and (b) the radars positioned to create a horizontal board [127]

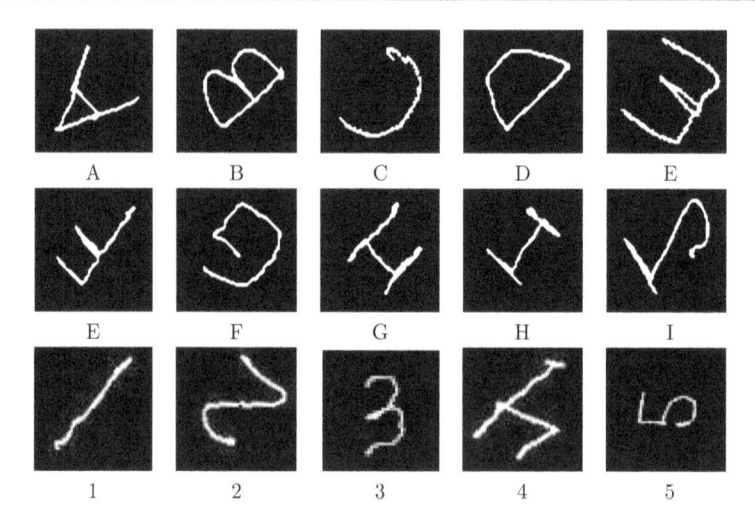

Figure 4.5 A few examples of 2D reconstructed trajectories from the dataset, illustrating the diversity in orientation, shape, and size of the trajectories, which adds complexity to the learning process

4.4 Air-writing With Sparse Network of Radars

In this section, we propose an air-writing system based on a sparse network of radars, specifically fewer than three, employing 1D deep convolutional-Long short-term memory-transposed-Convolutional (1D DCNN-LSTM-1D transposed DCNN) architecture. This architecture allows for the reconstruction and classification of the drawn character using only the range information from each radar. Real data collected using one and two 60-GHz FMCW radar sensors is utilized to demonstrate the effectiveness of our proposed air-writing solution. The content presented in this section aligns with our publication P1.

4.4.1 Signal Processing Chain

The proposed system only utilizes the range information from the radars. This range information is obtained with the signal processing step described in Sect. 4.1.1.1.

4.4.2 Proposed System

In this section, we present the mathematical framework underlying our proposed solution. Eq. 4.4 demonstrates that the trilateration problem has a unique solution. However, our approach involves utilizing fewer than three radars, leading to some reference points being unavailable. Consequently, Eq. 4.3 represents an undetermined system, with either an infinite number of solutions or no solution.

Assuming the utilization of a single radar, denoted as R_1, we can express the distance d_1 to radar at time t_n as follows:

$$(a(t_n) - a_1)^2 + (b(t_n) - b_1)^2 + (c(t_n) - c_1)^2 = d_1^2(t_n). \tag{4.9}$$

In the vicinity of the specified coordinates (a_0, b_0, c_0), the trajectory of the characters drawn in the air can be effectively approximated using either a constant velocity or constant acceleration model over a sequence of N consecutive measurements. Considering a constant velocity model with an unknown fixed velocity vector (v_a, v_b, v_c) in the vicinity of (a_0, b_0, c_0), we obtain:

$$\begin{cases} a(t_n) = a_0 + nv_a T \\ b(t_n) = b_0 + nv_b T \\ c(t_n) = c_0 + nv_c T \end{cases} \quad \text{for } n = 0, \ldots, N-1 \tag{4.10}$$

By applying this approximation to the measurements for $n = 0, \ldots, N-1$, the updated positional relationship at time $t_n = nT$ can be expressed as follows:

$$(a_0 + nv_a T - a_1)^2 + (b_0 + nv_b T - b_1)^2 + (c_0 + nv_c T - c_1)^2 = d_1^2(t_n) \tag{4.11}$$

Algebraic expansion of Equation (4.11) yields:

$$d^2(t_n) - d_1^2 = [1 \ -2a_1 \ -2na_1T \ -2b_1 \ -2nb_1T \ -2c_1 \ -2nc_1T \ t^2]\phi + g(\phi) \tag{4.12}$$

where $\phi = \begin{bmatrix} \rho_0^2 & a_0 & v_a & b_0 & v_b & c_0 & v_c & \dot{\rho}^2 \end{bmatrix}$ are the parameters to be estimated. Here, ρ_0 represents the radial distance of the target in the global reference, calculated as $\rho_0 = a^2 + b^2 + c^2$. Additionally, $\dot{\rho}_0^2$ denotes the radial velocity in the global reference, expressed as $\dot{\rho}_0^2 = v_a^2 + v_b^2 + v_c^2$.

The function $g(\phi)$ describes the non-linear relationship or interaction among the variables that need to be estimated. In the case of a constant velocity model, it represents the component of the velocity vector (v_a, v_b, v_c) that aligns with the direction of the radar position vector (a_0, b_0, c_0). However, for the constant acceleration model, the model captures more intricate and interactive relationships among the unknown variables.

Assuming the initial coordinates (a_0, b_0, c_0) are known is a reasonable assumption given that the trajectory is continuously drawn within the piece-wise approximation. However, for the first time instance, this assumption implies fixed starting coordinates, leading to an indeterminate offset from the reference. Nevertheless, this ambiguity does not affect the air-writing application significantly, as it relies on the relative positional trajectory. Equation (4.12) can be expressed in matrix-vector form as follows:

$$\tilde{a}(t_n) = \begin{bmatrix} -2na_1T & -2nb_1T & -2nc_1T & T^2 \end{bmatrix} \begin{bmatrix} v_a \\ v_b \\ v_c \\ \dot{\rho}^2 \end{bmatrix} + g\left(\begin{bmatrix} v_a \\ v_b \\ v_c \\ \dot{\rho}^2 \end{bmatrix}\right) \quad (4.13)$$

where the known terms have been incorporated into the $\tilde{a}(t_n)$ term. By combining N consecutive measurements and organizing them in matrix-vector notation, Equation (4.13) can be expressed as:

$$\begin{bmatrix} \tilde{a}(t_0) \\ \vdots \\ \tilde{a}(t_{N-1}) \end{bmatrix} = \begin{bmatrix} s(t_0)^T \\ \vdots \\ s(t_{N-1})^T \end{bmatrix} \tilde{\phi} + g(\tilde{\phi}) \quad (4.14)$$

The estimation of $\tilde{\phi}$ in Equation (4.14) presents a highly complex and non-linear problem due to several factors. These factors include the presence of the non-linear function $g(\tilde{\phi})$, an underlying unknown motion model, and the unknown number of measurements N until the piece-wise approximation remains valid. Unlike the linear relationship in Equation (4.3), this process cannot be solved through a simple mathematical formula.

Due to the absence of a closed-form expression for the approximation of $\tilde{\phi}$ and its highly dynamic nature with multiple hyperparameters, utilizing the short-term learning capability of LSTM (Long Short-Term Memory) can be advantageous for more effective learning. In addition to the local piece-wise approximation model, a global embedding is essential for the reconstruction of characters. This global embedding facilitates the reconstruction of trajectories using a fixed alphabet derived from the dictionary. The long-term global aspect can be represented as a dictionary learn-

4.4 Air-writing With Sparse Network of Radars

ing problem, wherein the main objective is to identify a dictionary that enables the learned embeddings of each character trajectory to possess a sparse representation.

Given that the input character trajectories are represented by $\mathbf{A} = [a_1, \cdots, a_k]$, where $a_i \in \mathbb{R}^d$, the objective is to obtain a dictionary $\mathbf{S} \in \mathbb{R}^{d \times n}$, where $\mathbf{S} = [s_1, \cdots, s_n]$ (allowing $n > d$, creating an overcomplete dictionary), and a representation $\phi = [\phi_1, \cdots, \phi_k]$, where $\phi_i \in \mathbb{R}^n$, such that $\|\mathbf{A} - \mathbf{S}\phi\|$ is minimized, while ensuring that the representation ϕ_i remains sufficiently sparse. Traditionally, this problem is formulated as:

$$L^*(A, S) = \min_{\phi \in \mathbb{R}^n} \|\mathbf{A} - \mathbf{S}\phi\|_2^2 + \lambda \|\phi\|_0 \tag{4.15}$$

The inclusion of the ℓ_0 norm in the aforementioned problem introduces nonconvexity, making the minimization task NP-hard. However, according to optimization and compressed sensing theory, it is widely recognized that utilizing the ℓ_1 norm in the minimization problem leads to a convex formulation, thereby rendering the problem tractable [132].

The ℓ_1 norm enforces sparsity on ϕ by allowing only a few non-zero coefficients, which are also referred to as atoms. The value of λ determines the balance between minimizing the error and achieving sparsity. Solving the aforementioned minimization problem enables the generation of a dictionary $\mathbf{S}$ that aims to closely approximate the majority, if not all, of the vectors from the input data $\mathbf{A}$. In our proposed system, the dictionary $\mathbf{S}$ is learned using LSTM's long-term learning capabilities.

Figure 4.6 presents a comparison between the conventional solution and the proposed solution. The functionalities of trilateration, $\alpha\beta$ tracking filter, and tracking history of the conventional pipeline are replaced with a 1D-DCNN for feature extraction, followed by LSTM and 1D transposed DCNN for temporal modeling of both local and global trajectories. The LSTM encodes the trajectory of each character that's been sketched into a condensed embedding feature vector. This feature vector acts as a comprehensive representation, encapsulating the vital traits of the trajectory. Following this, the encoded data is channeled into 1D transposed DCNN for the global reconstruction of the trajectory. In the conventional pipeline, classification is achieved using either a 2D-DCNN or LSTM.

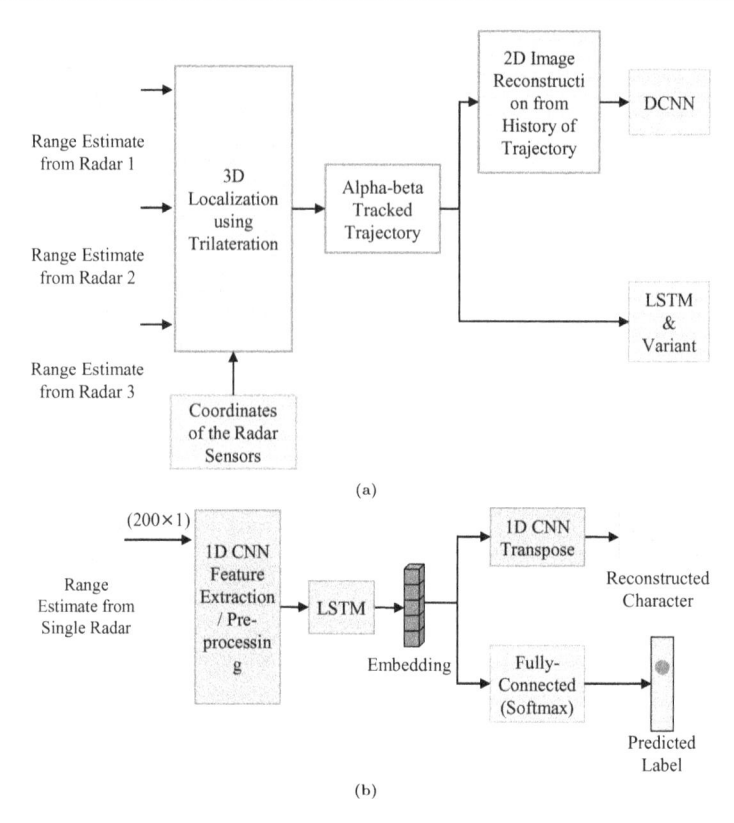

130   4 Radar-based Air-writing for Embedded Devices

Figure 4.6 Comparison between the conventional and proposed solutions for air-writing. (a) The conventional pipeline uses trilateration with three radars, $\alpha\beta$ tracking filter, and track history functionalities, with classification accomplished using 2D-DCNN or LSTM. (b) The proposed pipeline employs a 1D-DCNN for feature extraction, followed by LSTM for temporal modeling and 1D transposed DCNN for trajectory reconstruction, while classification is accomplished via a fully-connected softmax

4.4.3 Architecture & Learning

4.4.3.1 Architecture

The proposed 1D DCNN-LSTM-1D transposed DCNN architecture consists of four stages: feature extraction, encoding, classification, and reconstruction. During the feature extraction stage, the input trajectory is subjected to a 1D convolution operation to extract relevant and adequate features. The convolution operator's translation invariance property allows for modeling and extraction of local structures [133]. The features that are extracted in the feature extraction stage are subsequently fed into the encoding stage, where they are mapped to a condensed embedding using the LSTM layer.

In the proposed model, the LSTM layer plays a crucial role in compressing the input data and generating a learned representation, which serves as input for both the classification and reconstruction stages. Within the classification stage, the fully-connected layer utilizes a softmax function to calculate the conditional probability of the outputs, incorporating the learned features from the input trajectory. During the reconstruction or decoding stage, the compressed representation of the input trajectory is passed through multiple 1D convolution transpose layers. These layers aim to reconstruct the initial input trajectory as closely as possible.

Figure 4.7 illustrates the schematic of the architecture, which can be summarized as follows: Initially, an input array of size 200×1 (or 200×2 for two radars) is processed by a 1D CNN layer with 128 filters. This is followed by an LSTM layer with 100 filters. A dense layer with a softmax activation function is then applied to the output of the LSTM layer. The resulting output is subsequently passed through 10 1D CNN transpose layers. The first three layers employ 32 deconvolutional filters with a size of 7×1, while the subsequent three layers utilize 64 deconvolutional filters, each with a filter size of 7×1. Following that, the architecture comprises three subsequent layers, each consisting of 128 deconvolutional filters with a filter size of 7×1. The final layer of 1D CNN transpose consists of two deconvolutional filters of size 7×1.

Throughout all convolutional and transposed layers, the rectified linear unit (ReLU) activation function is employed. To mitigate overfitting, dropout has been implemented on the last layer of the 1D CNN transpose layers within the model.

Figure 4.7 Proposed 1D DCNN-LSTM-1D transposed DCNN network architecture where the architecture takes an input array of 200 × 1 (or 200 × 2 for two radars), consisting of a 1D CNN layer with 128 filters, followed by an LSTM layer with 100 filters. A dense layer with softmax activation is appended to the LSTM output, which is then fed into 10 1D CNN transpose layers. The architecture includes 32 deconvolutional filters of size 7 × 1 in the first three transpose layers, followed by 64 filters of the same size in the next three layers. The last three transpose layers each contain 128 filters of size 7 × 1. The final layer has 2 deconvolutional filters of size 7 × 1. The ReLU activation function is used for all convolutional and transposed layers. Dropout has been applied to the last layer of 1D CNN transpose layers to prevent overfitting

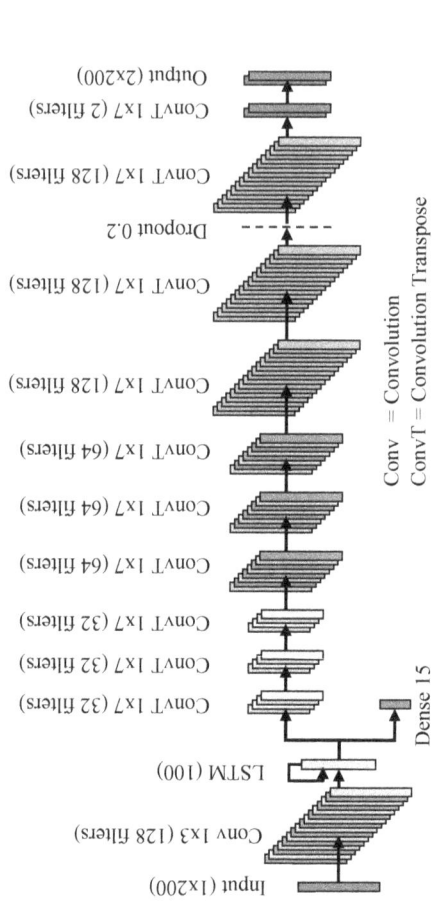

4.4.3.2 Loss Function

In the classification stage, a softmax classifier is employed to calculate the probabilities of the class labels, providing insights into the model's confidence for each class. The softmax function employs cross-entropy as a loss function, which can be represented as:

$$L_{CE} = -\frac{1}{L}\sum_{k=1}^{K}\sum_{l=1}^{L} y_l^k \log(h_\theta(x_l, k)), \quad (4.16)$$

where the symbol L denotes the total number of training examples, while K represents the number of classes. The notation y_l^k refers to the target label for training example l and class k. The variable x represents the input for training example m, and h represents the model with weights θ.

In the reconstruction stage, Mean Squared Error (L_{MSE}) is employed to quantify the dissimilarity between the input and reconstructed trajectories. This function is minimized throughout the training process and is mathematically expressed as:

$$L_{MSE} = \frac{1}{L}\sum_{l=1}^{L}(h_\theta(x^l) - y^l)^2 \quad (4.17)$$

where x represents the input trajectory, while y represents the reconstructed trajectory. The overall loss is calculated as a weighted sum of the classification loss (L_{CE}) and the Mean Squared Error (L_{MSE}) using the following expression:

$$L_t = w_1 L_{CE} + w_2 L_{MSE} \quad (4.18)$$

The loss weights w_1 and w_2 are set to 100 and 0.20 respectively.

4.4.3.3 Weight Initialization & Learning Schedule

The weights of the 1D convolutional layers were initialized using Xavier initialization [111], while the dense layer weights were initialized by sampling from a normal distribution with mean 0 and standard deviation of 0.01. For the biases of the dense layer, samples were drawn from a normal distribution with a mean of 0.5.

The Adam [109] optimizer was employed as the learning criterion, dynamically calculating the learning rate for each weight during the learning process. The adaptive learning rate is computed using estimates of the first moment ($\beta_1 = 0.9$) and the second moment ($\beta_2 = 0.999$) of the gradients.

4.4.4 Results & Discussion

4.4.4.1 Dataset
For the proposed approach, We have used the dataset explained in Sect. 4.3.

4.4.4.2 Reconstruction Results
To perform a quantitative analysis of the reconstruction results, we computed the Mean Squared Error (L_{MSE}) between the reconstructed trajectories and the reference trajectories. The L_{MSE} results obtained from our system can be found in Table 4.2. Initially, we calculated the L_{MSE} for each individual sample, and subsequently, the average of these values was computed across all the samples.

Table 4.2 The average mean squared error (L_{MSE}) for the reconstruction

Radar configuration	Avg. L_{MSE} (in e-4)
Single radar	6.8 ± 1.5
Two radars	1.4 ± 0.7

Figure 4.8 exhibits several examples of reconstructed trajectories. The blue trajectory represents the reference trajectory, while the red trajectory corresponds to the reconstructed trajectory using two radars, and the green trajectory represents the reconstructed trajectory using a single radar. The reference trajectories were generated utilizing trilateration, and the best trajectory was chosen through visual inspection. As depicted in Fig. 4.8, it is evident that our system is proficient in reconstructing the original trajectories in both scenarios, regardless of whether a single radar or two radars were employed.

4.4 Air-writing With Sparse Network of Radars

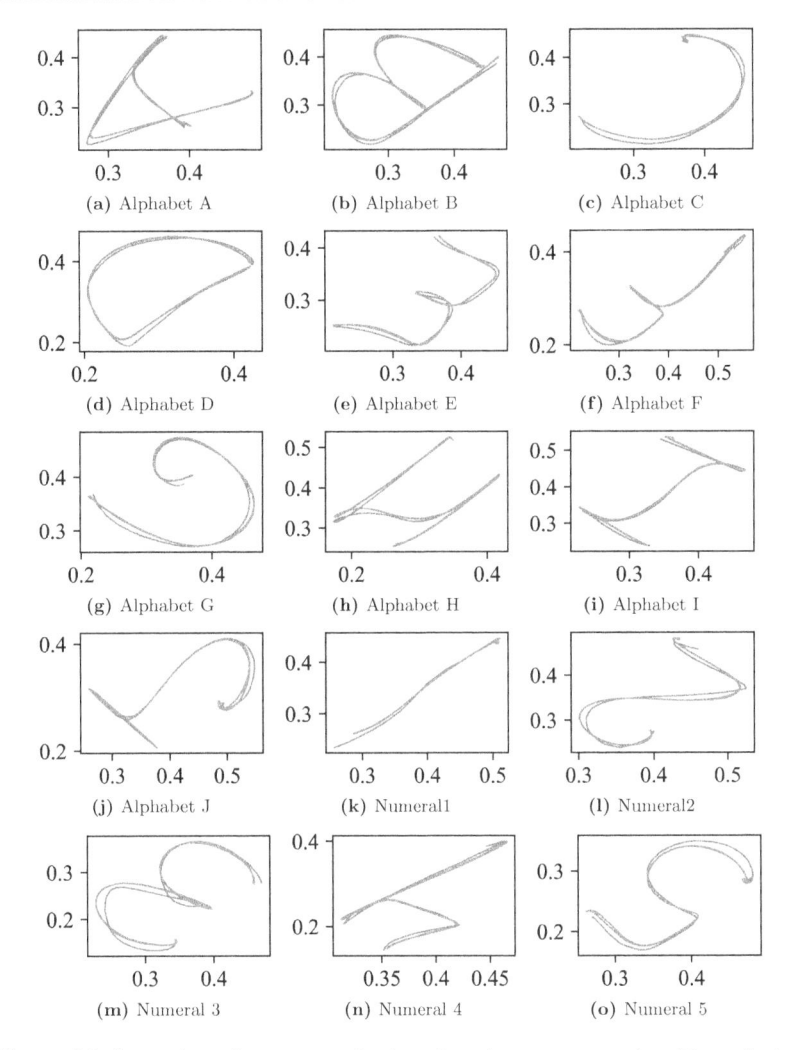

Figure 4.8 Comparison of reconstructed trajectories using one or two radars. Blue trajectory represents the reference trajectory, red represents the reconstructed trajectory using two radars, and green represents the reconstructed trajectory using a single radar. The system is capable of accurately reconstructing the original trajectories in both scenarios

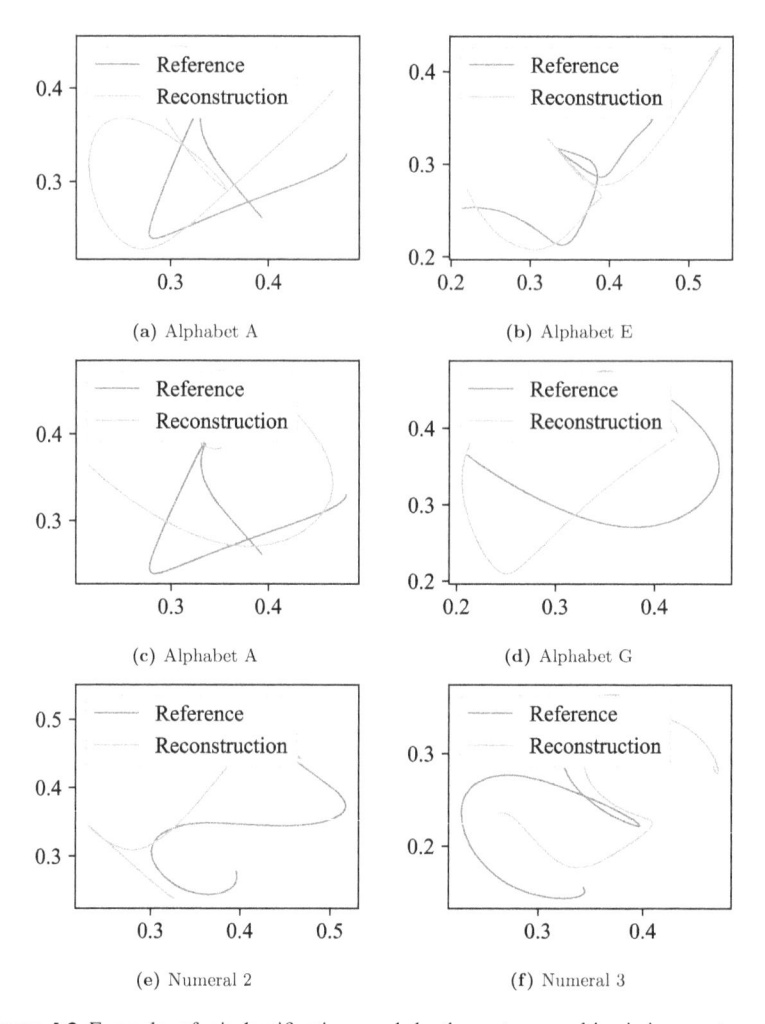

136   4 Radar-based Air-writing for Embedded Devices

Figure 4.9 Examples of misclassifications made by the system, resulting in incorrect reconstructed trajectories. Reconstructed trajectories (in orange) do not correspond to the actual characters, such as characters A, B, and G. The learned dictionary of characters is utilized to generate these reconstructions. Misclassifications by Dense 15 correspond to incorrect trajectory reconstruction by the 1D transposed DCNN. The misclassification is likely caused by an incorrect projection of the input trajectory feature into the wrong cluster by the learned LSTM

4.4.4.3 Classification Results

To assess the classification outcomes, we employed accuracy as the evaluation metric. The average classification results for both single radar and two radar setups are presented in Table 4.3. Notably, the proposed approach outperforms the baseline methods when two radars are utilized, achieving an accuracy of 97.33% ± 2.67%, compared to the highest baseline accuracy of 98.33%. In the case of a single radar scenario, our proposed approach achieves an accuracy of 90.33% ± 4.44%.

Table 4.3 Comparison of classification accuracy: proposed SNN model versus other baseline models

Approach	No. of radars	Trajectory	Accuracy (%)
Baseline LSTM-CTC [127]	3	Global	93.33
Baseline BLSTM-CTC [127]	3	Global	96.67
Baseline ConvLSTM-CTC [127]	3	Global	98.33
Baseline DCNN [127]	3	Global	98.33
Proposed model_1	1	Local	**90.33 ± 4.44**
Proposed model_1	2	Local	**97.33 ± 2.67**

4.4.4.4 Discussion

In our proposed radar-based air-writing system, we conducted a comprehensive investigation into the capabilities of deep neural networks. Our results indicate that deep neural networks can effectively learn implicit features directly from the range information provided by the radar. This eliminates the requirement for pre-processing steps like trajectory extraction or trilateration, as observed in previous studies [127]. Furthermore, unlike trilateration which requires a minimum of three radars, our proposed method enables global trajectory reconstruction using only one or two radars. The utilization of the LSTM temporal model allows us to leverage the trajectory's correlation with its previous measurements, effectively converting the generally under-determined problem into a well-determined one. Moreover, our architecture is compact, with a size of approximately 5.49 Mb, making it highly efficient for implementation on low-cost hardware. As shown in Table 4.3, our proposed method achieves a comparable level of accuracy to the current state-of-the-art approach [127] by utilizing only two radars. Specifically, our method achieves an accuracy of 97.33% ± 2.67%, while the state-of-the-art approach achieves an accuracy of 98.33%. This improvement can be attributed to the replacement of the trilateration block for feature extraction with 1D convolution, demonstrating the effectiveness of this alternative approach.

The trilateration algorithm is prone to inaccuracies due to finger occlusion on one or more radars and the possibility of missed target detection on any radar, leading to unreliable trajectory coordinates. Moreover, manually crafted features obtained through trilateration or extended Kalman filter require efficient preprocessing and data augmentation techniques to handle different orientations and starting points effectively. The proposed system overcomes these inaccuracies by inherently learning the feature images, thus mitigating the issues associated with finger occlusion and missed detections. However, when only one radar is employed, the accuracy of the system decreases by 6 − 9%. This decrease can be attributed to the limited dimensions available to the model for extracting relevant features from the trajectory, in comparison to the configurations with two radars. One potential solution to overcome this limitation is the utilization of a faster frame-rate radar system.

To further evaluate the performance of our system, we conduct a visualization of the high-dimensional feature space. The LSTM model maps the input feature vector from the 1D DCNN into hidden features with a dimensionality of 100. In order to assess the distinctiveness of these feature spaces for each class, we employ the t-Distributed Stochastic Neighbour Embedding algorithm (t-SNE) to project these features along with their respective labels.

The visualizations presented in Figs. 4.10(a) and 4.10(b) showcase a 2D feature representation obtained through t-SNE for the single and two radar configurations, respectively. Each class is distinguished by a unique color, resulting in a total of 15 clusters. Each point within the visualizations corresponds to a feature. It is evident from the t-SNE embeddings of the test data that distinct clusters are formed, indicating the capability of the proposed system to effectively separate different classes.

Furthermore, for the two radar configurations, the projection demonstrates a remarkably accurate representation. Nonetheless, there are cases in which the test sample trajectory of certain characters is erroneously projected into a cluster associated with a different character. This scenario is demonstrated in Fig. 4.10(a), where samples from characters A and G are inaccurately projected into the cluster assigned to character D. As a consequence, the classifier yields misclassifications, and the 1D transposed DCNN generates erroneous reconstructions. Similarly, clusters 2 and 1 are observed to be grouped together due to the resemblance in their trajectories. However, the clustering performance notably improves when utilizing two radars (Fig. 4.10(b)) where only characters D, B, G, and H have few incorrect samples projected into a cluster associated with a different character.

4.4 Air-writing With Sparse Network of Radars

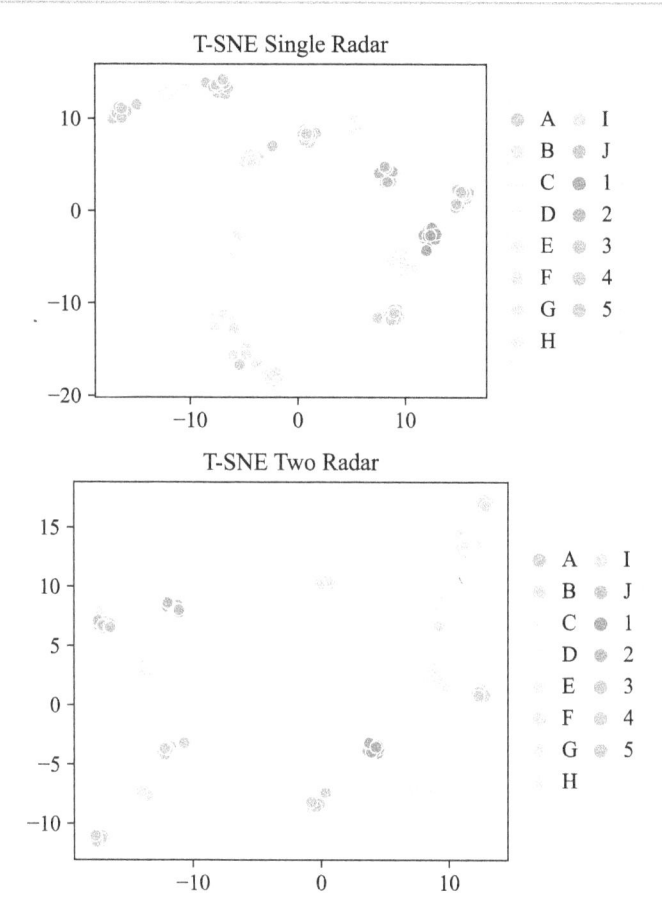

Figure 4.10 Visualization of 2D feature representations for single and two radar configurations using t-SNE. Each class is represented by a different color, with a total of 15 clusters depicted. However, misprojections are observed for characters A and G, which are erroneously placed in the cluster of character D. Similarly, clusters 2 and I are grouped together. These misclassifications lead to errors in reconstructions by the 1D transposed DCNN, as shown in (a). Despite these challenges, the two radar configuration demonstrates improved clustering, shown in (b), with only a few misclassified samples projected into a different character's cluster. This indicates the effectiveness of utilizing two radars for enhanced feature representation and reduced misclassifications

In the case of the reconstruction process, the proposed system successfully recovers the original trajectories from the learned embedding space, as illustrated in Fig. 4.8. The reference trajectories, depicted in blue color, symbolize the optimal trajectories obtained through trilateration by utilizing three radars. Each radar measures the distance of the marker within its field of view, and this information is utilized by the trilateration algorithm to estimate the marker's position. The output of the algorithm provides a localized and smoothed global trajectory, representing the marker's movement relative to the reference coordinates.

In contrast, our proposed solution utilizes the LSTM-learned embedding to reconstruct corresponding global trajectories using the 1D transposed DCNN. From Fig. 4.8, it is evident that our proposed system achieves highly accurate reconstruction of the original trajectories, regardless of whether two radars (red) or a single radar (green) is employed. The green reconstruction displays only a minor deviation when compared to the red reconstruction. This deviation is also reflected in the L_{MSE} values listed in Table 4.2, where the L_{MSE} increases in the case of a single radar.

Figure 4.9 displays several examples of misclassifications made by the system, resulting in reconstructed trajectories corresponding to a different character. For instance, for characters A, B, and G, the system produces reconstructed trajectories (indicated by the orange color) that do not correspond to the actual characters. Similarly, for characters E, G, 2, and 3, the resulting reconstructed trajectories are F, D, I, and 5, respectively. These reconstructions are generated by utilizing the learned dictionary of characters, as described in Sect. 4.4.2. In the majority of cases, it has been noticed that the misclassification of a character (by Dense 15) aligns with an incorrect reconstruction of the trajectory by the 1D transposed DCNN. This indicates that the erroneous projection of the input trajectory feature into the incorrect cluster by the learned LSTM is the root cause of the misclassification. Nevertheless, despite its simplicity, the proposed model demonstrates robustness in real-time recognition and reconstruction.

4.4.5 Conclusion

In this section, we present a novel air-writing system that utilizes a sparse network of FMCW radars, specifically fewer than three. Our proposed system relies exclusively on the range information acquired from these radars, which is then processed through a 1D DCNN-LSTM-1D transposed DCNN architecture. This architecture enables the reconstruction of the 2D drawn image relative to a reference point and facilitates the classification of the drawn character. We provide comprehensive evidence of the effectiveness of our proposed air-writing solution, evaluating its performance in terms of both reconstruction and classification using a sparse network of FMCW radars.

4.5 Radar Trajectory-based Air-writing Recognition

In this section, we propose a novel approach to air-writing that focuses on using only one or two radars to sense the local hand trajectory. To recognize the drawn character based on this local trajectory, we employ a 1D temporal convolutional network (TCN), explained in Sect. 2.2.4, that performs simultaneous feature extraction and temporal modeling. The proposed end-to-end solution achieves an average accuracy of approximately 99.11% and 91.33% for the two-radar and one-radar-based approaches, respectively, demonstrating superior performance compared to the proposed solution in Sect. 4.4 and state-of-the-art methods. Moreover, the proposed system enables continuous writing of words containing 2 to 4 characters. The content presented in this section aligns with our publication P13.

4.5.1 Signal Processing Chain

The proposed approach only employs the range of data provided by the radars. These range estimates are acquired by means of the signal processing procedure outlined in Sect. 4.1.1.1.

4.5.2 Proposed Solution

In our proposed approach, we utilize a sparse network of fewer than three radars, resulting in an underdetermined trilateration problem as described in Equation (4.4) of Sect. 4.1.1.2. This underdetermined system may have either no solution or an infinite number of solutions. By utilizing deep learning models, we exploit the temporal information of range changes to approximate the global data required for precise classification of the drawn characters. In doing so, we extensively harness the capabilities of neural networks, enabling them to learn features directly from the range and range change data, even in scenarios where the system is underdetermined.

Assume that the local coordinate trajectory $\rho_1(t)$ represents the range at time t for a specific character drawing, and let $\phi(t)$ denote the global coordinate trajectory $\bigl(\phi_a(t), \phi_b(t), \phi_c(t)\bigr)$. We introduce a smooth function $f(.)$ that maps or transforms the local trajectory $\rho_1(t)$ to the global trajectory $\phi(t)$. Employing a second-order Taylor series expansion centered at $t = 0$, we can approximate the global trajectory in the following manner:

$$\phi(t) = f(\rho_1(t)) = f(\rho_1(0)) + f'(\rho_1(0))(\rho_1(t) - \rho_1(0)) \\ + \frac{f''(\rho_1(0))}{2}(\rho_1(t) - \rho_1(0))^2, \quad (4.19)$$

where,

$f'(\rho_1(0)) = [\frac{\partial \phi_a(t)}{\partial \rho_1(t)}, \frac{\partial \phi_b(t)}{\partial \rho_1(t)}, \frac{\partial \phi_c(t)}{\partial \rho_1(t)}]_{t=0}$ and $f''(\rho_1(0)) = [\frac{\partial^2 \phi_a(t)}{\partial \rho_1^2(t)}, \frac{\partial^2 \phi_b(t)}{\partial \rho_1^2(t)}, \frac{\partial^2 \phi_c(t)}{\partial \rho_1^2(t)}]_{t=0}$. The variable $t \in \tau_i$ where τ_i is the ith time segment over which the Taylor expansion is valid. It is important to note that when τ_i is degenerate, the function $f(.)$ does not have a unique solution. We can express the multi-class classification problem as follows:

$$c_p = g_\theta \bigg(\{\phi(t \in \tau_1), \phi(t \in \tau_2), \cdots, \phi(t \in T)\} \bigg), \quad (4.20)$$

where c_p represents the predicted drawn character, g_θ denotes the neural network function with parameters θ, and T represents the number of time segments utilized to approximately piece-wise the entire trajectory. These segments are fixed but unknown, denoted as $f'i(\rho_1(0)), f''i(\rho_1(0))i = i^{i=T}$. In this approach, we propose the utilization of TCN to perform the operation $g_\theta(.)$. We then compare the obtained results with solutions that rely on LSTM and 1D CNN-LSTM.

Figure 4.11 provides a visual comparison between the conventional solution and the proposed solution. The conventional pipeline involves trilateration, $\alpha\beta$ tracking filter, and track history functionalities. However, in the proposed solution, these components are substituted with 1D TCN for the temporal modeling of local trajectories. The TCN effectively extracts features by modeling the temporal aspects of each character, and the classification is carried out using a fully-connected softmax layer. On the other hand, in the conventional pipeline, classification is accomplished through 2D-DCNN or LSTM methods.

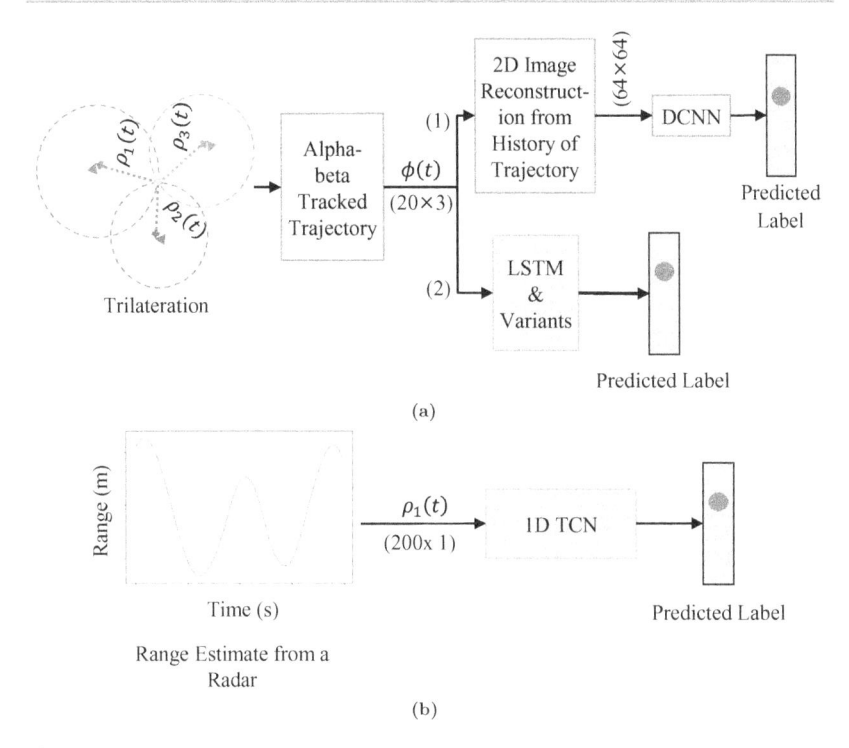

4.5 Radar Trajectory-based Air-writing Recognition

Figure 4.11 Comparison between the conventional pipeline shown in (a) and proposed solution shown in (b). The conventional pipeline involves trilateration using 3 radars, $\alpha\beta$ tracking filter, and track history functionalities. The proposed solution replaces these with 1D TCN for temporal modeling of local trajectories. The TCN extracts features through temporal modeling for each character, and classification is performed using fully-connected softmax. In contrast, the classification in the conventional pipeline is accomplished through 2D-DCNN or LSTM using global trajectory $\phi(t)$. The local trajectories coordinates are represented by $\rho_1(t)$, $\rho_2(t)$ and $\rho_3(t)$ of the three radars respectively

4.5.3 Architecture & Learning

The input to the architecture is a 200 × 1 vector, where 200 represents the time steps and 1 represents the feature (which becomes 2 in case of two radar sensors). Each sensor's reading contributes to one channel, and these features are passed to the TCN. The TCN combines features from different time steps and stores them in a single feature vector using dilation factors that increase exponentially. In this proposed setup, only one range value (or two in the case of two radars) or sensor reading is fed to the TCN at each time step. The TCN comprises of nine layers of dilated causal convolution, where the dilation factors increase as $[2^0, 2^1, \cdots, 2^8]$. For each layer, 20 filters with a kernel size of 6 and a stride of 1 are used. During training, weight normalization, ReLU activation, and dropout with a rate of 0.3 are added after each dilated causal convolution layer. The output layer of the network is a fully connected layer appended with a softmax function.

4.5.3.1 Loss Function

We employed the widely-used cross-entropy loss function for multi-class classification tasks, which can be mathematically represented as:

$$L(\hat{X}_i, X_i) = -\sum_{j=1}^{c} x_{ik} \log(p_{ij}), \quad (4.21)$$

where X_i is a one-hot encoded target vector $(x_{i1}, \cdots, x_{ic})$, and $\hat{x}_i = (\hat{x}i1, \cdots, \hat{x}ic)$ represents the output of the classifier. The term x_{ij} is equal to 1 if the ith element is in class j, otherwise, it is 0. The term p_{ij} denotes the probability of class label j being the correct classification for the ith element.

4.5.4 Results & Discussion

4.5.4.1 Dataset

For training and evaluation, we have used the dataset explained in Sect. 4.3.

4.5.4.2 Classification

To evaluate the classification performance of our proposed system, we utilized average accuracy as a metric. Table 4.4 presents the average accuracy results for both the single radar and two radar configurations. The findings demonstrate that our proposed method with two radars achieves a similar level of accuracy compared to the state-of-the-art trilateration-based approach [127] and to the proposed model (Proposed model_1) discussed in Sect. 4.4.2.

Table 4.4 [Comparsion of the classification accuracy achieved by the proposed SN model with other baseline models

No. of radars	Trajectory	Approach	Accuracy (%)	Memory/Parameters
ConvLSTM [127]	3	Global	98.33	1.56 MB / 133, 903
DCNN [127]	3	Global	98.33	751 kB / 173, 631
LSTM	1	Local	89.32 ± 6.66	1.10 MB / 93, 465
LSTM	2	Local	93.33 ± 5.32	1.10 MB / 94, 065
1D CNN-LSTM	1	Local	90.33 ± 4.44	891 kB / 94, 139
1D CNN-LSTM	2	Local	97.33 ± 2.67	891 kB / 94, 139
Proposed model_1	1	Local	90.33 ± 4.44	NA
Proposed model_1	2	Local	97.33 ± 2.67	NA
Proposed model_2	1	Local	91.33 ± 4.66	643 kB / 41, 635
Proposed model_2	2	Local	99.11 ± 0.89	644 kB / 41, 775

4.5.4.3 Discussion

We propose an air-writing system that leverages the use of one or two radar systems and employs a TCN model for feature extraction and temporal modeling of the radar trajectory. The system eliminates the need for additional preprocessing steps and directly learns implicit features from the trajectory data.

The proposed approach has advantages for space-limited devices like personal monitors or screens and provides a cost-effective solution. This is achieved by utilizing TCN for simultaneous feature extraction and temporal modeling of the trajectory in relation to its prior measurements. By incorporating temporal modeling, the proposed system effectively addresses the challenge of an under-determined problem. Moreover, unlike the system presented in [127], which focuses solely on single character recognition, our proposed system has the capability to detect words and continuous numerals. This expansion in functionality allows for more versatile and comprehensive air-writing applications.

To validate the effectiveness of the proposed method, we conducted a comparative analysis with other deep temporal models, including LSTM and 1D CNN-LSTM. The evaluation results demonstrate that our proposed system achieves superior classification accuracy compared to these models. Furthermore, our proposed architecture exhibits a compact memory footprint (approximately 644 kB without considering quantization, pruning, and fusion) and requires a smaller number of trainable parameters (only 41, 635) compared to alternative models. These characteristics make our proposed system an efficient and practical solution for deployment on commodity hardware.

As shown in Table 4.4, the proposed system achieves a comparable level of accuracy, specifically 99.11%±0.89%, to the state-of-the-art method presented in [127], which achieves an accuracy of 98.33% in classification. This improvement can be attributed, in part, to the replacement of the hand-crafted feature extraction method (trilateration) utilized in prior works. Hand-crafted features often suffer from inaccuracies caused by occlusion of fingers on one or more radars and missed target detection on a specific radar, leading to unreliable global coordinates. The proposed approach effectively addresses the aforementioned inaccuracies by learning the features and temporal modeling jointly within the model.

As observed in Table 4.4, the accuracy slightly decreases by 4 − 8% when using a single radar, which can be attributed to the reduced dimensions available for the model to learn the required features. To mitigate this limitation, increasing the framerate of the radar system or incorporating Doppler trajectory information could be considered. These enhancements would provide the model with more informative data, thereby improving the accuracy of the system.

Figure 4.12(a) illustrates the change in range over frames for selected letters and numbers captured by both radars. The range data from each radar is represented by the blue and red lines, respectively. In Fig. 4.12(b), the corresponding global coordinates of the characters and numbers are presented. These global coordinates are reconstructed using the range data from three radars and the trilateration technique.

The results of continuous character writing are showcased in Fig. 4.13. The characters are drawn sequentially with varying pauses between them, depending on the user's interaction. The thresholding method, as mentioned earlier, was employed to identify the start and end points of each character during the drawing process over time. As shown in Fig. 4.13, two words, BAD and FACE, are presented as examples of continuous writing with the separation between each word indicated. The black dotted horizontal line represents the minimum range threshold, which signifies when the target is within the field of view (FoV) or virtual board. The range values obtained from the radar are higher when the target is inside the FoV compared to when it is outside. The thresholding technique used was simple yet highly effective, achieving a segmentation accuracy of 100% when evaluated on over 50 words with character lengths ranging from 2 to 4.

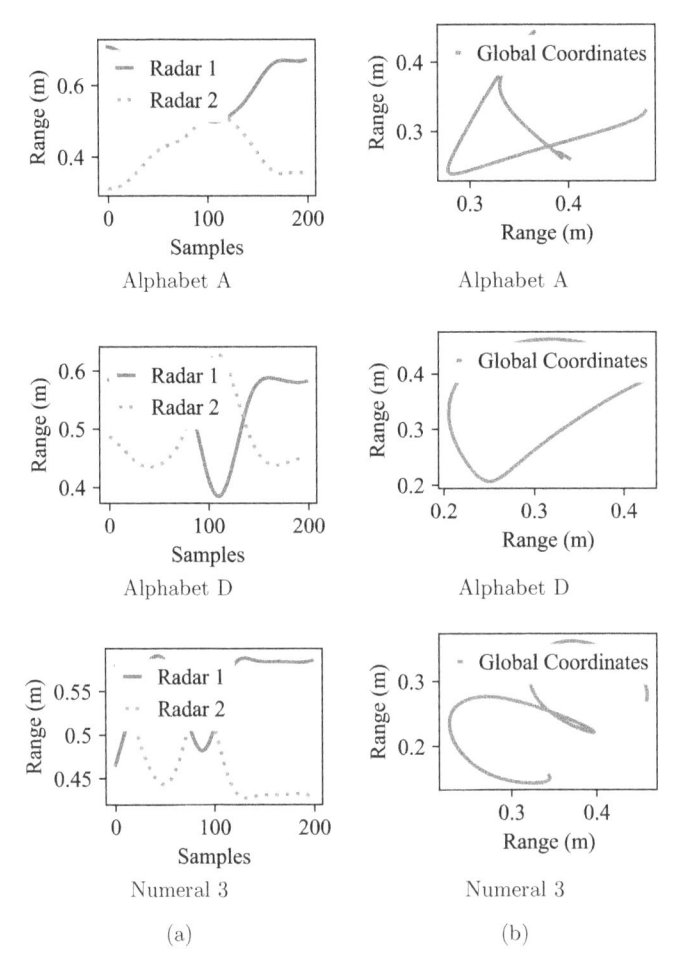

4.5 Radar Trajectory-based Air-writing Recognition   147

Figure 4.12 Comparison of range changes over frames for characters and numerals from two radars, represented by blue and red lines in column (a). The global coordinates of the characters and numerals are also presented in column (b), obtained by using range data from three radars and trilateration to reconstruct the global trajectory. In the case of a single radar, only the blue curve reading is fed to the system, while in the case of two radars both the blue and red curves readings are fed to the system

(a) Word BAD

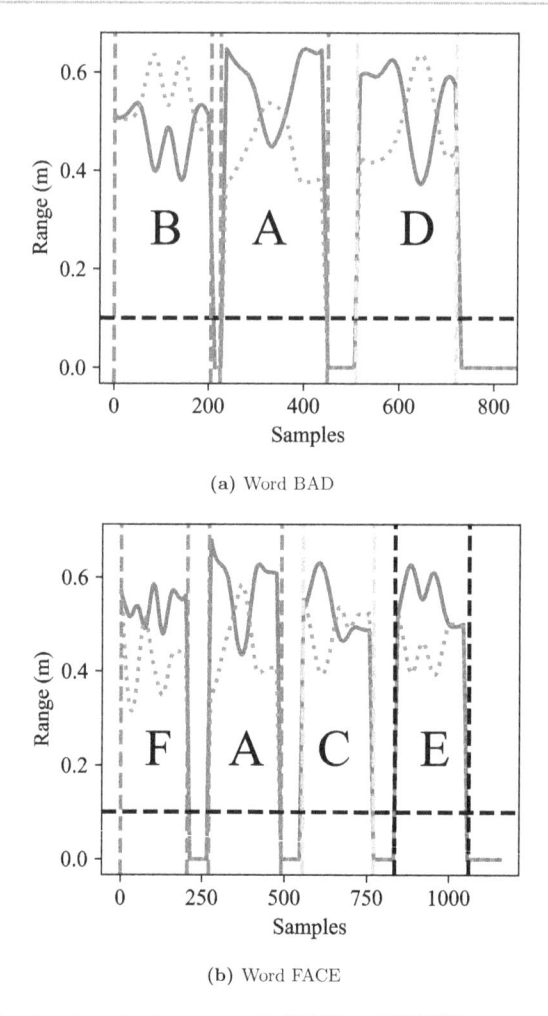

(b) Word FACE

Figure 4.13 Local trajectories for two words "BAD" and "FACE" are presented in (a) and (b), respectively, with the separation of each word indicated. The black dotted horizontal line represents the minimum range threshold established when the target is within the field of view (FoV) or virtual board. The simple thresholding technique, evaluated on over 50 words with character lengths varying from 2 to 4, achieved 100% segmentation accuracy. For the single radar configuration, only the blue curve is fed to the system, while for the two radar configuration, both the red and blue curves are fed to the system

4.5.5 Conclusion

To address the practical limitations associated with traditional air-writing systems that require three or more radars, we propose a novel approach that leverages the local range trajectory obtained from one or two radars to recognize characters drawn on a virtual board. Our solution involves utilizing a 1D TCN classifier model, which effectively captures the temporal patterns of the local trajectory and performs accurate character classification. We compare our proposed solution with other deep-Nets and trilateration-based approaches, in terms of their classification accuracy and memory footprint. Furthermore, the proposed system facilitates the uninterrupted composition of words consisting of 2 to 4 characters.

4.6 Radar-based Air-Writing Using Spiking Neural Networks

In contrast to the existing literature that mainly focuses on deep learning methods for air-writing that can pose significant computational and energy demands, making them unsuitable for energy-efficient edge IoT devices, in this section we propose a highly energy-efficient air-writing system that employs SNNs. The SNN utilizes fine-range estimates and trilateration from the radar network to recognize and classify the trajectory of the written character. Our proposed system achieves a comparable level of classification accuracy, reaching 98.6%, when compared to state-of-the-art deep learning methods. Furthermore, the proposed SNN model exhibits memory efficiency, with a compact size of 3.7MB, making it storage-friendly. We validate the effectiveness of our proposed approach in real-time using a network of 60-GHz FMCW radar chipsets. The content presented in this section aligns with our publication P8.

4.6.1 Signal Processing Chain

We used the same signal processing techniques as outlined in Sect. 4.1.1 and depicted in Fig. 4.14(a). Range estimates were obtained using the method described in Sect. 4.1.1.1. Trilateration, as explained in Sect. 4.1.1.2, was used to track and localize the target, and generate the 3D trajectory of the air-written character. The trajectory was then smoothed using an alpha-beta ($\alpha\beta$) filter, and a 2D image was reconstructed from the trajectory for input into our SNN model for classification.

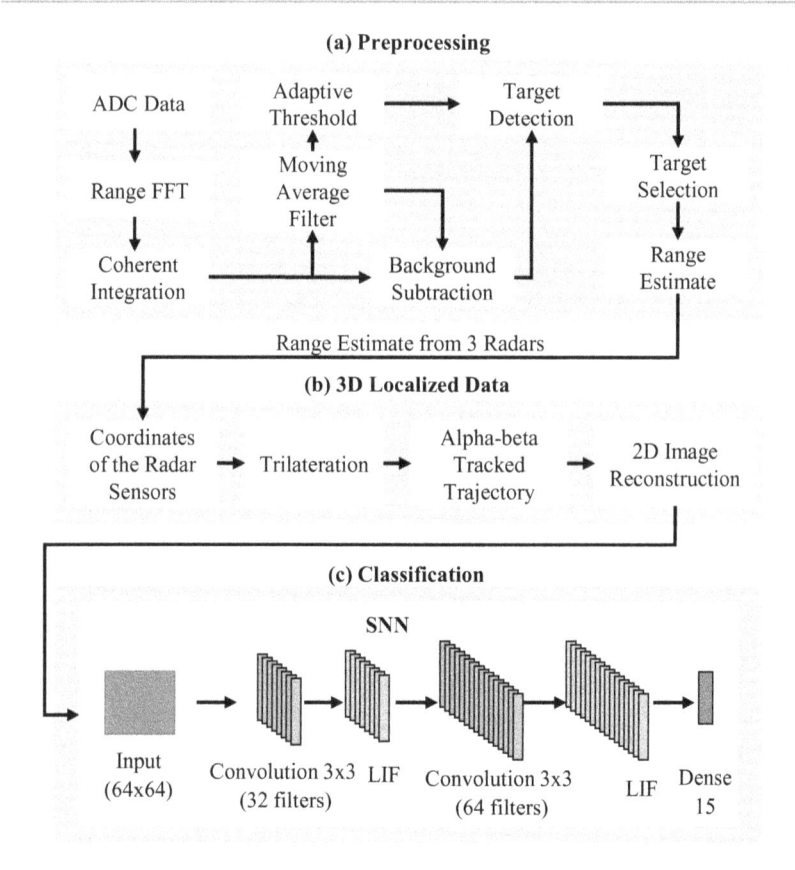

Figure 4.14 Coloumn (a) displays the signal processing steps involved in obtaining range information from FMCW radar. Spectral analysis is performed along each chirp to obtain time delay estimates, which are then combined using pulse integration. An MTI filter is applied to filter out stationary objects, and target detection and selection are performed through thresholding. Once range information from three radars is available, trilateration (shown in (b)) is used to determine the global 3D coordinates with respect to a reference. (c) shows the architecture of the proposed SNN. The network takes in 64 × 64 2D images and comprises of two convolutional layers with 32 and 64 kernels, respectively and the output layer has 15 nodes. SoftLIF is used as an activation function for all the layers in training. During testing, the SoftLIF neurons are replaced with spiking LIF, and the trained biases and weights are used to construct the network. The classification is achieved using a softmax classifier, which provides the classification probabilities. The presentation time, which represents the time duration for capturing the spiking output, is set to 25

4.6.2 Architecture & Learning

The architecture of the proposed SNN is depicted in Fig. 4.14(c) and comprises convolutional layers followed by dense layers. The network was constructed using the nengoDL simulator, which provides a differential approximation of biologically plausible neurons, such as LIF [134]. This approximation allows for backpropagation training of the network.

The network takes as input the 2D images generated from the preprocessing steps, which have a size of 64 × 64. Following the input layer are two convolutional layers with a kernel size of 3 and a total of 32 and 64 kernels respectively. LIF is used as an activation function for both convolutional layers. The final layer is a dense layer with 15 nodes, which serves as the output layer. The objective function used to train the network is multi-class cross-entropy, and the activation function used is SoftLIF, which is an approximation of LIF.

To classify the input, we utilized the softmax classifier, which provides classification probabilities. The loss function used by the softmax classifier is cross-entropy, and can be represented mathematically as follows:

$$\mathcal{L} = -\frac{1}{K} \sum_{c=1}^{C} \sum_{k=1}^{K} x_k^c \log(\eta_\theta (i_k, c)), \quad (4.22)$$

where x_k^c denotes the true label for training example k for class c, and η is the model with weights θ that takes i as input for the training example.

Following the training phase using SoftLIF, the SoftLIF neurons are replaced with spiking LIF neurons during the testing phase, and the trained biases and weights are employed to construct the network. This ensures that the testing model is a fully functional SNN. To capture the spiking output over time, the test inputs are repeated multiple times, a process known as presentation time. In our case, we empirically set the presentation time to 25, considering the trade-off between accuracy and latency.

4.6.3 Results & Discussion

4.6.3.1 Dataset
For training and evaluation of our proposed setup, We have used the dataset explained in Sect. 4.3.

4.6.3.2 Data Augmentation

To improve the model's generalization capability, we employed data augmentation techniques on the training samples. This involved creating transformed versions of the data by applying random scaling, random rotation, random translation, and random skewing. By introducing these variations, we aimed to increase the diversity of the training data and enhance the model's ability to handle different input configurations and variations in real-world scenarios.

4.6.3.3 Hyperparameters

As with deepNets, determining the optimal hyperparameter settings is crucial for the learning performance of SNNs. In Table 4.5, we present the optimal hyperparameter values for our model. These values were obtained through a grid search

Table 4.5 The hyperparameters used for training the proposed SNN model

Type	Parameters	Value	Definition
Model Hyperparameters	Learning rate (η)	0.001	The step size of the optimizer at each iteration while moving towards a minimum of a loss function
	Output synaptic delay	10 ms	The latency of the output for a given input
	Type of neuron model	LIF	Leaky integrate and fire model
	Batch size	100	The number of samples processed before the model is updated
	No. of epochs	10	Total number of iterations
	Time steps during the testing	25	The repetition of the input during the testing phase
	No. convolutional layers	2	The total number of layers used in the model
	No. of Filters	32, 64	The total number of convolution filters used. The first layer uses 32 and 2nd layer use 64.
Neuron Parameters	τ_{rc}	0.02	Time constant for Membrane RC (in seconds). This parameter determines how fast the membrane voltage decays to zero without input.
	τ_{ref}	0.002	Time It is the Absolute refractory period in seconds. Following a spike, it determines how long the membrane voltage remains zero.
	min voltage	0	Stands for the minimum membrane voltage.
	amplitude	0.01	It is the scaling factor on the neuron output.

process, where we systematically explored different combinations of hyperparameters while monitoring the SNN network's performance in terms of overfitting and underfitting. The selected hyperparameters represent the configuration that yielded the best balance between model complexity and generalization performance.

4.6.3.4 Performance Evaluation

The performance of the proposed method was evaluated by measuring the average classification accuracy. The results presented in Table 4.6 demonstrate that the proposed SNN method achieves a comparable level of accuracy (98.6%) to the state-of-the-art approaches, including the proposed model (Proposed model_1) in Sect. 4.4.2 and the proposed model (Proposed model_2) in Sect. 4.5.2. Notably, the proposed SNN achieves this level of accuracy while utilizing a smaller memory size of approximately 3.7 MB, making it an efficient option in terms of storage requirements. The accuracy measurements were obtained through multiple trials, with 20% of the dataset randomly selected as the test set for each trial.

Table 4.6 Comparison of classification accuracy between the proposed solution and state-of-the-art methods

No. of radars	Trajectory	Approach	Accuracy (%)	Memory/Parameters
ConvLSTM [127]	3	Global	98.33	1.56 MB / 133, 903
DCNN [127]	3	Global	98.33	751 kB / 173, 631
LSTM	1	Local	89.32 ± 6.60	1.10 MB / 93, 465
LSTM	2	Local	93.33 ± 5.32	1.10 MB / 94, 065
1D CNN-LSTM	1	Local	90.33 ± 4.44	891 kB / 94, 139
1D CNN-LSTM	2	Local	97.33 ± 2.67	891 kB / 94, 139
Proposed model_1	1	Local	**90.33 ± 4.44**	NA
Proposed model_1	2	Local	**97.33 ± 2.67**	NA
Proposed model_2	1	Local	**91.33 ± 4.66**	643 kB / 41, 635
Proposed mode_2	2	Local	**99.11 ± 0.89**	644 kB / 41, 775
Proposed model_3	3	Global	**98.6**	NA

4.6.3.5 Estimated Energy Consumption

To estimate the energy consumption of our proposed system, we utilized the hardware metrics of the μBrain chip, as defined in [108]. The energy consumption per classification was calculated using the following formula:

$$E_c = N_s \times E_s + \delta T \times P_l \tag{4.23}$$

where E_c denotes the energy consumed per classification, N_s represents the maximum number of spikes during the classification process, E_s (equal to 2.1 pJ) stands for the energy per spike, P_l (equal to 73 µW) represents the static leakage power and δT signifies the inference time. Assuming the inference time is 28 ms, the energy consumption per classification for the proposed system can be calculated as $E_c = 2.13$ µJ.

To compare the energy efficiency of SNN hardware with other deep learning hardware, a benchmarking study was conducted in [113] using a keyword spotting application. The benchmarking involved commercially available hardware devices, and the dynamic energy cost per inference was measured across these devices. Fig. 4.16 illustrates the results of this benchmarking, showing the difference between the total energy consumption during one inference and the energy consumed while the hardware is idle [113]. In their experiments, a single inference entailed passing a feature vector through a feed-forward neural network with two hidden layers to predict a probability distribution over alphabetical characters. The results demonstrate that the SNN hardware achieved a significant improvement of 10× in power efficiency.

The firing patterns of the proposed SNN model for six different characters and numerals (2, H, I, A, 5, and C) are presented in Fig. 4.15. The model starts firing with the correct class label after a few time steps, and the probability of correct classification increases with increasing time steps. This is due to the integration of spikes over a longer time period, which improves the accuracy of the model. The proposed system is a relatively simple solution, yet it facilitates robust real-time recognition of both alphabets and numerals while maintaining low power consumption.

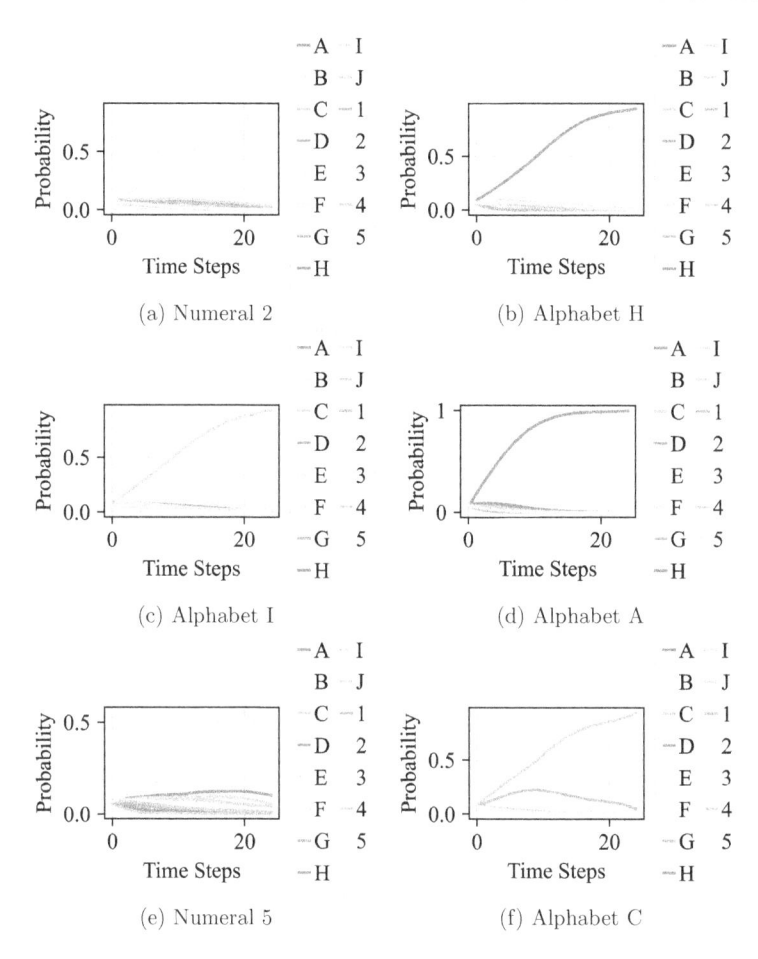

4.6 Radar-based Air-Writing Using Spiking Neural Networks

(a) Numeral 2

(b) Alphabet H

(c) Alphabet I

(d) Alphabet A

(e) Numeral 5

(f) Alphabet C

Figure 4.15 The firing patterns of the proposed SNN model for six different characters and numerals (2, H, I, A, 5, and C). The model starts firing for the correct class after a few time steps, and the accuracy of the model improves with longer integration of spikes over time

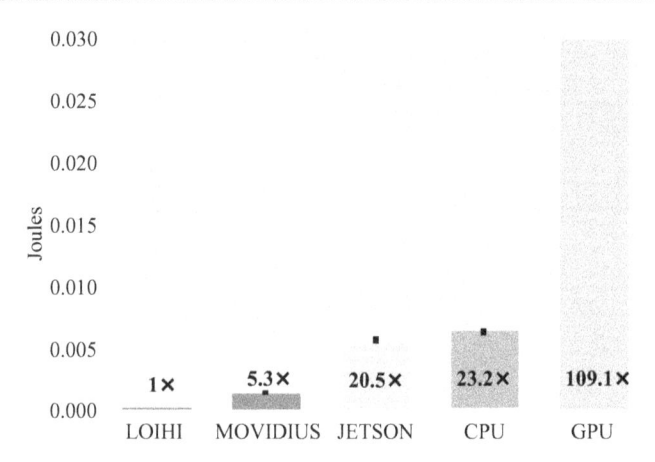

156   4   Radar-based Air-writing for Embedded Devices

Figure 4.16 Energy efficiency comparison of different hardware for keyword spotting application [113]. The dynamic energy cost per inference is measured using commercially available accelerators. SNN hardware demonstrates up to a 10x improvement in power efficiency compared to other deep learning hardware

4.6.4   Conclusion

We propose a new method for recognizing air-written characters using a combination of radar networks and SNN. The hand marker is localized and tracked using range estimates obtained from the network radars and trilateration technique. The motion trajectory, represented as a 2D image, is then processed by the SNN for character recognition. Our proposed system achieves a high accuracy level (98.6 %) while using a small memory size (3.7 MB) and has superior energy efficiency due to the use of SNN compared to other state-of-the-art methods.

4.7   Radar-based Air-Writing System Using Spiking Legendre Memory Unit

This section introduces an innovative radar-based air-writing system that utilizes a SNN to enable users to input characters into a user interface by drawing on an imaginary board. To optimize the SNN for the specific application, we propose the use of a Genetic Algorithm (GA) to determine the optimal parameters of the Spiking Legendre Memory Unit (SLMU). Our proposed solution achieves a similar level of accuracy, with 98.53% for the two-radar scenario, while surpassing the performance of deep learning methods for single radar with an accuracy of 95.37%. Moreover, the estimated energy consumption of our solution is significantly lower, measuring at 2.04 µJ, in contrast to deepNets counterparts that consume energy in the order of mJ. Additionally, our proposed systems require less storage memory, with 490kB for a single radar and 564.2kB for two radars, demonstrating smaller memory footprints compared to existing state-of-the-art solutions. The content presented in this section aligns with our publication P5.

4.7.1 Signal Processing Chain

The proposed system only utilizes the range information from the radars. This range information is obtained with the signal processing step described in Sect. 4.1.1.1.

4.7.2 Architecture & Learning

The proposed approach combines GA and SNN models, where the GA is used to find optimal parameters during training and the LMU is trained based on the parameters identified by GA. The GA model uses the accuracy of the LMU as the objective function. The optimization of the LMU training parameters was achieved using GA, which involved creating a population of potential solutions, known as individuals, as shown in Fig. 4.17. The GA evaluates the potential solutions based on a fitness function, which in our case aims to maximize the accuracy of the LMU. In each generation, the GA performs three steps: (1) selects the best individuals, (2) discards the individuals with poor fitness values, and (3) generates new individuals through crossover and mutation to replace the discarded ones. This iterative process continues until the GA outputs the best individual that yields the highest accuracy for the LMU. Table 4.8 displays the GA-optimized parameters. The chromosome, which represents the candidate solutions, was represented in binary form where each gene is denoted as either 0 or 1.

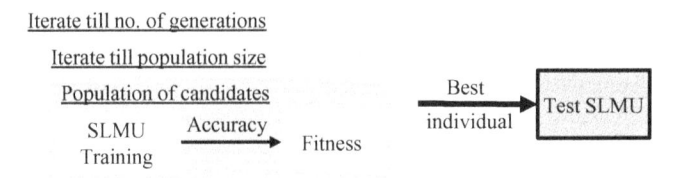

```
Iterate till no. of generations
  Iterate till population size
    Population of candidates                Best
      SLMU    Accuracy                   individual    Test SLMU
     Training  ────────►   Fitness
```

Figure 4.17 The flow of the Genetic Algorithm (GA) and Spiking Legendre Memory Unit (SLMU) hybrid model shows the optimization process using GA to find the optimal parameters for training the SLMU, which is evaluated based on the accuracy of the SLMU. The GA operates in a population-based manner where new individuals are created through crossover and mutation, and the fitness function is used to select the best individuals for each generation. The process continues until the GA outputs the best individual that yields the highest accuracy for the SLMU

4.7.2.1 Architecture

The proposed architecture is shown Fig. 4.18, which is based on LMUs. The network is constructed using nengo-dl[38] and the network is initially trained using non-spiking *tanh* as the activation function in a traditional backpropagation approach. When constructing an LMU network, careful selection of parameters such as units, order, and theta is crucial. Units determine the number of neurons used, while order represents the number of Legendre polynomials used for the orthogonal representation of the sliding window, which captures changes in the input signal. Increasing the order enables the representation of faster changes in the input. Theta determines the amount of temporal information stored in the LMU. The values of units, order, and theta obtained using GA were 225, 191, and 5 for a single radar, and 221, 220, and 25 for two radars.

The LMU output is then fed to a dense layer of 15 neurons representing the number of classes in the dataset.

A softmax classifier is applied to the output ooposed SNN architecture f this layer using cross-entropy as the metric, which is defined as:

$$L_{CE} = -\frac{1}{K} \sum_{m=1}^{M} \sum_{k=1}^{K} z_k^m \log\left(H_\theta\left(x_k, m\right)\right), \tag{4.24}$$

where K is the total number of training samples and M is the number of classes. The true label of the kth sample for the mth class is represented by z_k^m. The model weights θ trained on input x are denoted by H.

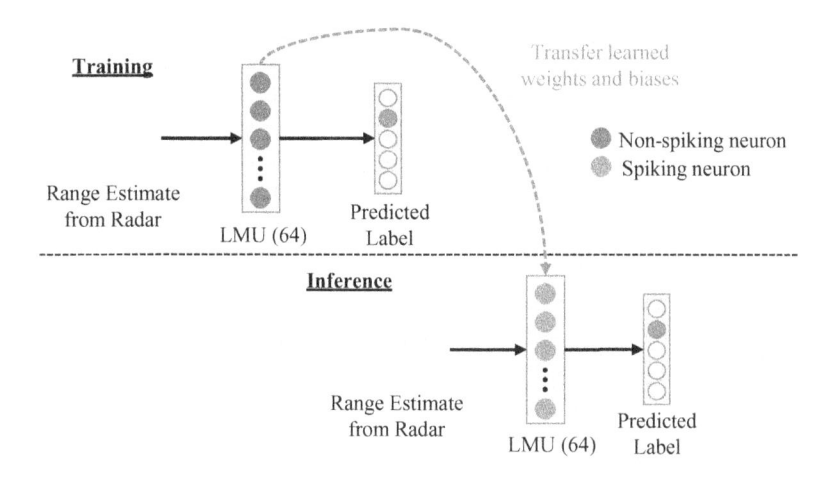

4.7 Radar-based Air-Writing System Using Spiking Legendre Memory Unit

Figure 4.18 The prflow, involves training the model with rate-based (non-spiking) neurons using a conventional backpropagation technique. During inference, the non-spiking neurons are replaced with spiking neurons in a new model. The weights and biases learned during training are then transferred to the test model

The learning schedule used in this work is adaptive moment estimation (Adam), with a learning rate of 0.001. The decay rates for the 1st and 2nd moment are set to 0.9 and 0.999, respectively, while epsilon (ϵ) is set to $1e^{-7}$ for numerical stability.

4.7.2.2 Model Training and Testing

During the initial training phase, the model is trained using rate-based neurons with backpropagation. After training, the weights and biases of the trained model are saved. During the testing phase, a new model is created with spiking neurons, and the saved weights and biases are used to connect the spiking neurons, thus converting the model to a spiking model.

4.7.3 Results & Discussion

4.7.3.1 Dataset

For training and evaluation of our proposed setup, we have used the dataset explained in Sect. 4.3.

4.7.3.2 Classification Performance

The performance of the proposed system was evaluated by comparing its accuracy with other state-of-the-art methods. As shown in Table 4.7, the proposed method achieves a similar level of accuracy as the state-of-the-art methods with two radars and outperforms them in the case of a single radar. Moreover, it achieves similar level of performance as that of the proposed model (Proposed model_1) in Sect. 4.4.2, the proposed model (Proposed model_2) in Sect. 4.5.2 and the proposed model (Proposed model_3) in Sect. 4.6.2.

Table 4.7 The classification accuracy achieved by the proposed solution in comparison to the performance of state-of-the-art methods

Approach	No. of radars	Trajectory	Accuracy (%)	Memory/Parameters
ConvLSTM [127]	3	Global	98.33	1.56 MB / 133, 903
DCNN [127]	3	Global	98.33	751 kB / 173, 631
LSTM	1	Local	89.32 ± 6.66	1.10 MB / 93, 465
LSTM	2	Local	93.33 ± 5.32	1.10 MB / 94, 065
1D CNN-LSTM	1	Local	90.33 ± 4.44	891 kB / 94, 139
1D CNN-LSTM	2	Local	97.33 ± 2.67	891 kB / 94, 139
Proposed model_1	1	Local	**90.33 ± 4.44**	**5.49 MB / 469, 277**
Proposed model_1	2	Local	**97.33 ± 2.67**	**5.49 MB / 469, 277**
Proposed model_2	1	Local	**91.33 ± 4.66**	**643 kB / 41, 635**
Proposed mode_2	2	Local	**99.11 ± 0.89**	**644 kB / 41, 775**
Proposed model_3	3	Global	**98.6**	**NA**
Proposed model_4	1	Local	**95.37 ± 1.63**	**490 kB**
Proposed model_4	2	Local	**98.53 ± 1.47**	**564.2 kB**

4.7.3.3 Discussion

We introduce a novel approach to air-writing, which utilizes a single radar and SNN. Our proposed method does not require any pre-processed features or global trajectory reconstruction, such as trilateration used in [127]. Instead, the SNN learns the features directly from the range data. Compared to [127], our system is more robust against miss detection since it does not rely on data from multiple radars. Additionally, our system is based on SNN which are highly energy efficient and hardware-friendly, making them ideal for edge applications. In Table 4.7, we evaluated the classification accuracy of the proposed system against other state-of-the-art systems. The results show that when using two radars, the proposed system achieves a similar

4.7 Radar-based Air-Writing System Using Spiking Legendre Memory Unit

accuracy level of 98.53%, while when using a single radar, it outperforms them with an accuracy of 95.37%. Table 4.8 demonstrates that the optimal hyperparameters provided by GA were used to obtain these results. The system's performance can be attributed to the temporal aspects of the SNN. Additionally, the proposed model has a small size (490 kB for a single radar and 564.2 kB for two radars), which requires a small memory footprint. The assistance of GA in fine-tuning the model parameters aided in achieving a low memory footprint.

Table 4.8 Optimal parameters for SLMU obtained using the Genetic Algorithm

Hyperparameter	Single radar	Two radars	Definition
Type of Neuron	Spiking Tanh	Spiking Tanh	Neuron model used
τ_{ref}	1	1	Absolute refractory period expressed in seconds
LMU's units	225	221	Number of employed neurons
LMU's order	191	220	Number of Legendre polynomials
LMU's theta	5	25	Amount of time kept in the LMU's internal memory
Learning rate	0.002	0.003	Learning rate of our Adam optimiser
Minibatch Size	32	64	Number of concurrent inputs fed to the network
Epochs	2000	2000	Number of passes of the entire training dataset
Synapse	0.2 s	0.1 s	Output synaptic delay
Time steps	1	1	Test input representation time or time steps during the testing

The optimal parameters for the SLMU were obtained by conducting Binary GA simulations for 25 generations with a population size of 100. Parent selection was performed using a Steady State Selection criteria, where some individuals with high fitness were selected for new offspring and some individuals with low fitness were removed and replaced with new offspring in each generation. For the next generation, a two-point crossover was used to combine two individuals and genetic diversity was introduced through scramble mutations.

In Fig. 4.19, the accuracy evolution with generations is displayed for single radar. It is noticeable that the accuracy improves as the number of generations increases. Towards the end of the generations, optimal parameters are generated by the GA, resulting in a high accuracy rate.

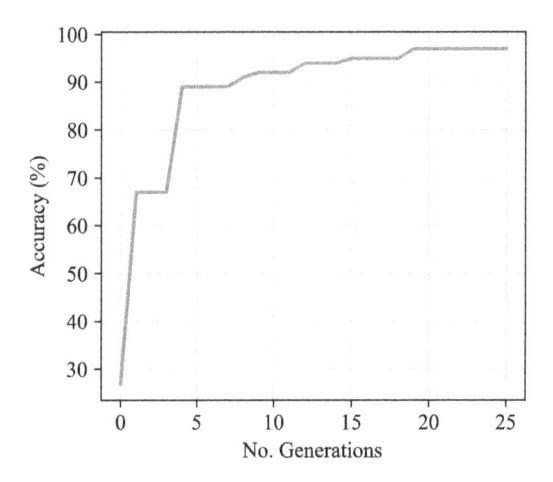

Figure 4.19 Evolution of accuracy with respect to generations for single radar. The graph shows an improvement in accuracy as the number of generations increases. The optimal parameters generated by GA at the end of the generations lead to a high accuracy rate

Figure 4.20 displays some trajectories of the characters and numerals from the dataset, where the range estimates obtained from radar 1 and radar 2 are represented by red and green lines, respectively. When using a single radar system, only the green line is used for input, while both the green and red lines are used for input when using two radars.

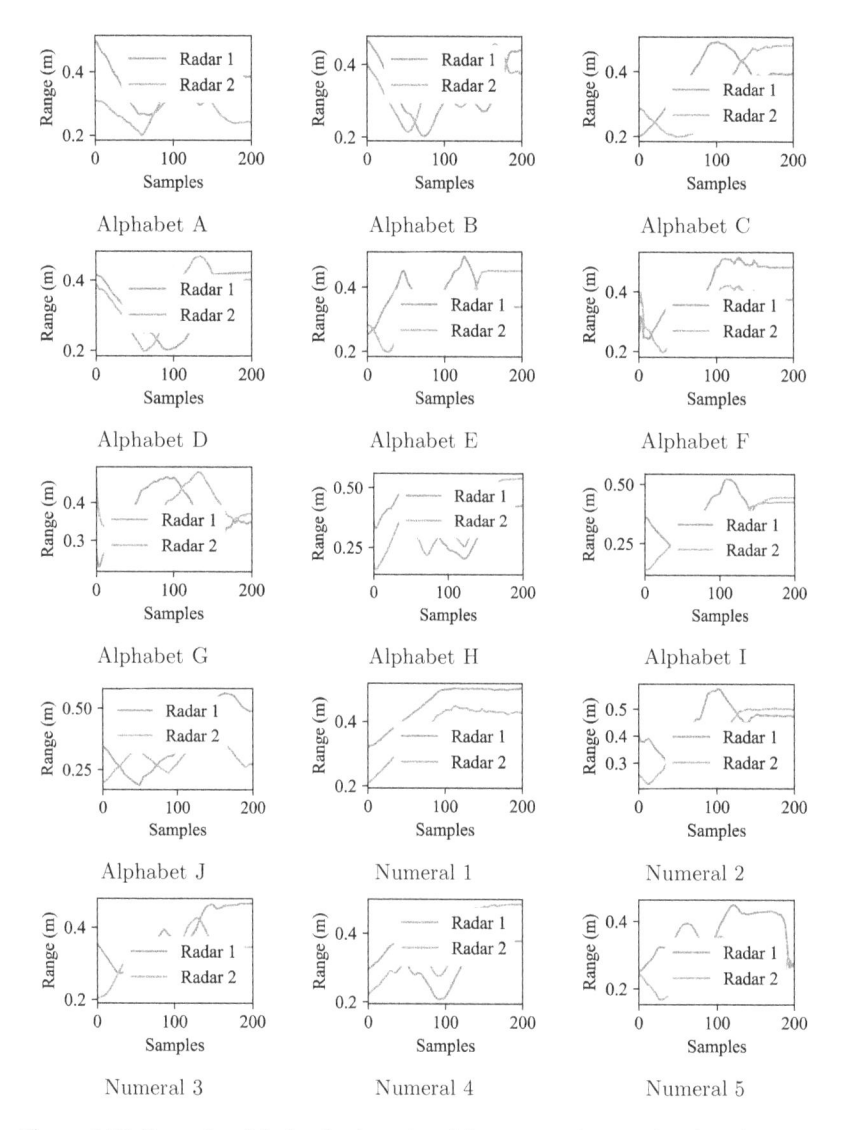

4.7 Radar-based Air-Writing System Using Spiking Legendre Memory Unit 163

Figure 4.20 Examples of the local trajectories of characters and numerals, where the system receives only the green curve reading when using one radar and both the green and red curve readings when using two radars

4.7.3.4 Performance Evaluation

To estimate the energy consumption, the hardware metric of the μBrain chip [108] can be used. The energy consumption per classification is mathematically defined as follows:

$$EC = ES \times NS + \delta T \times PL, \qquad (4.25)$$

where EC represents the energy consumed per classification, ES denotes the energy per spike (2.1 pJ), NS is the maximum number of spikes generated during classification, δT is the inference time, and PL is the static leakage power (73 µW). Assuming an inference time of 28 ms, the proposed system's energy consumption per classification is 2.04 µJ. This is significantly lower than deep neural networks counterparts that consume energy on the order of mJ [113].

4.7.4 Conclusion

This section proposes a novel approach to single radar air-writing using spiking neural networks that greatly enhance the energy efficiency of the system. Our proposed Spiking Legendre Memory Unit (SLMU) model utilizes a genetic algorithm to optimize its hyperparameters. The model takes only the range information obtained by the radar for a marker trajectory. Our proposed model performs on par with our proposed approaches in Sect. 4.4 and 4.5 for two radars and surpasses them for a single radar while consuming minimal power and occupying a small memory footprint. We believe that this work will facilitate the practical deployment of air-writing systems for portable or edge devices.

Time Series Forecasting of Healthcare Data 5

5.1 Introduction

The use of health tracking devices has grown significantly over the last 30 years [135]. These gadgets are utilized by clinicians to monitor the health of their patients and by individuals for self-tracking purposes.

Keeping track of one's diet and health habits can be a beneficial tool for achieving fitness and well-being goals, as it allows users to observe their progress over time and effectively work towards their desired outcomes. As an illustration, an athlete getting ready for a competition might establish nutritional goals and find it advantageous to track the anticipated progress of their calorie consumption over the coming weeks by using past data. As health data is of a sensitive nature, it is crucial to ensure privacy is upheld when handling it. The General Data Protection Regulation (GDPR) in the European Union mandates that systems for processing sensitive data provide strong guarantees for data privacy and security to prevent misuse of sensitive health care data [136]. To ensure privacy while performing time-series forecasting, we utilize federated learning (FL) [46], [47] and differential privacy (DP) techniques [137]. FL enables forecasting models to operate on edge devices, without transferring sensitive data to a central server. Rather than data, only model training parameters are exchanged with central models. Additionally, we investigate energy-efficient learning models for forecasting that can function on edge devices with restricted computing power. Although conventional deepNets have been successful in time-series forecasting applications, they tend to be inefficient in terms of energy consumption. The bulk of energy consumption in deepNets is attributed to the multiply-accumulate (MAC) operations between layers [93]. As a result, the research community has primarily concentrated on minimizing MACs by utilizing techniques such as smaller networks, weight quantization, and pruning [86], [87].

© The Author(s), under exclusive license to Springer Fachmedien Wiesbaden GmbH, part of Springer Nature 2024
M. Arsalan, *Optimization of Spiking Neural Networks for Radar Applications*,
https://doi.org/10.1007/978-3-658-45318-3_5

This chapter only covers the realistic synthetic data generation and the time series forecasting component of the state-of-the-art forecasting pipeline. For the privacy aspect, readers can refer to the study in [48].

5.1.1 State-of-the-Art

The state-of-the-art aligns with the publication P17. In the initial stage, individual user data records are created by aggregating raw data points. The central coordinator utilizes updates received from clients to enhance a global model, which is defined by a specific set of hyperparameters. Both the client and server employ the Adam optimizer, while the Standard Federated Averaging algorithm [47] is used for aggregation. To strike a balance between model convergence and computational efficiency, a random sample comprising 10% of users is selected for each round of Federated Learning (FL) [47].

Subsequently, the records are transmitted to the clustering mechanism, where a streaming k-means algorithm, combined with pattern-matching techniques, is applied to identify users with similar characteristics. The resulting cluster patterns, which represent sequences of centroid IDs, are stored by the central coordinator in FL, rather than the aggregated raw data records.

Based on these cluster patterns, the server maintains k federated models, where k corresponds to the number of groups. A visual representation of the process for a randomly selected client in one communication round is illustrated in Fig. 5.1. In step 1, the client named $client_p$ sends its updates. The coordinator then locates $client_p$ and determines that it belongs to cluster $k - 1$ (step 2). Consequently, the coordinator retrieves the model associated with cluster $k - 1$ ($model_k - 1$) and updates it using the received information from $client_p$ (step 3). Finally, the updated model is shared with the client.

In the noisy learning approach, TensorFlow Privacy mechanisms are utilized to add noise to the weights and transmit them to the federated aggregator. Both the baseline and clustered scenarios involve clients adding noise to their true update. The "clip" and "noise multiplier" parameters, as described in [138], control the amount of noise added. The standard deviation of the added noise is calculated by multiplying these two parameters. Furthermore, the aggregated user records are sent separately to the clustering mechanism, as depicted in Fig. 5.2.

In the noisy data setup, the FL process follows a similar pattern to the standard process, with the modification that clients introduce noise to their local datasets before improving the federated model. Consequently, the weights of each model sent as updates to the coordinator inherently include noise. The process is illustrated

5.1 Introduction 167

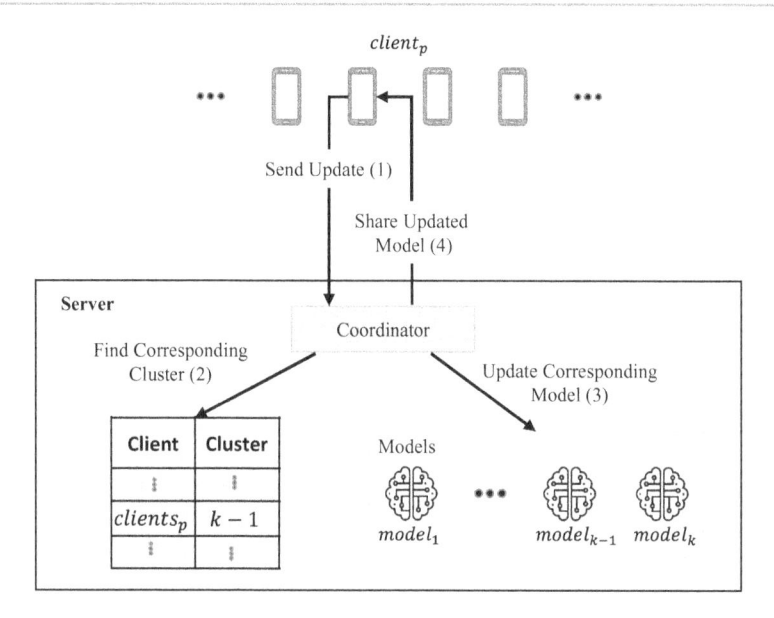

Figure 5.1 Visualization of communication rounds in the Federated Learning (FL) process with clustering, assumed to be grouped into k clusters

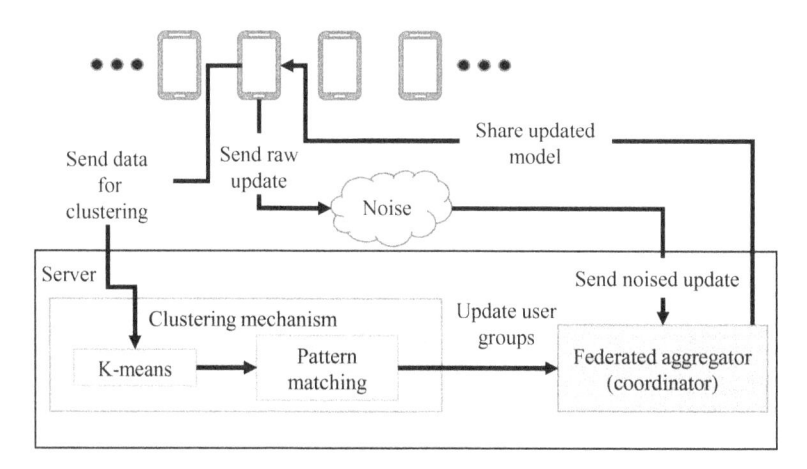

Figure 5.2 Noisy learning: clustered FL using streaming k-means. The baseline model is a traditional FL setup (not shown separately)

in Figs. 5.3 and 5.4. It should be noted that the clustering mechanism operates on aggregated data records that have been noised, which may affect the clustering quality. Furthermore, the prediction is conducted locally and on clean data.

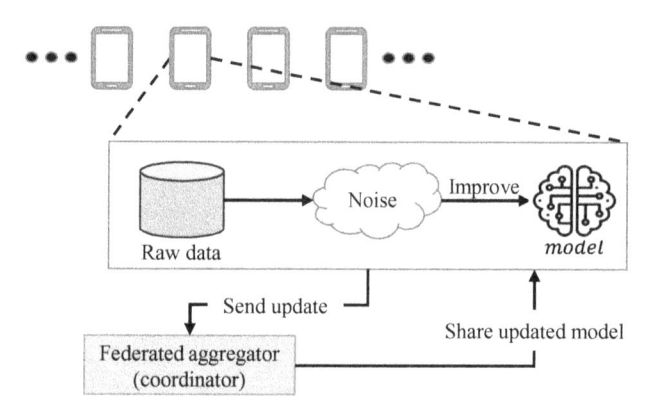

Figure 5.3 Noisy data: a) Baseline FL

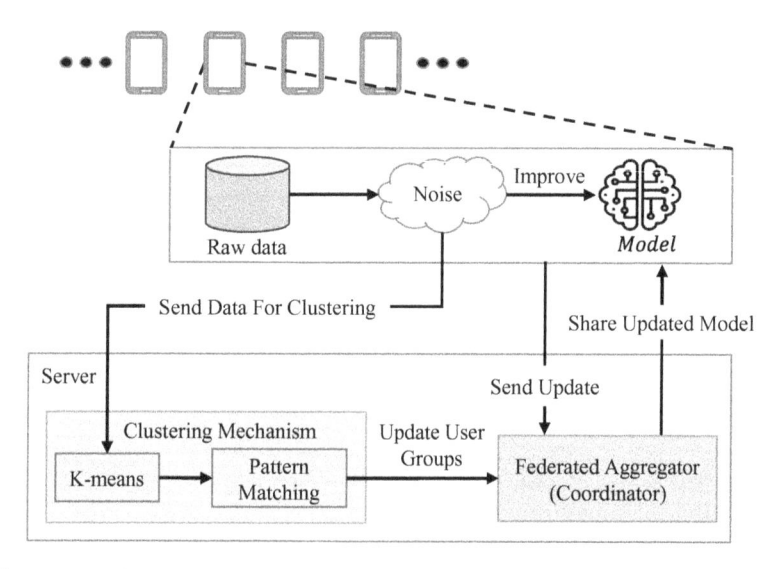

Figure 5.4 Noisy data: b) Clustered FL using streaming k-means

Previous research has indicated that an epsilon value greater than 1 may not provide adequate privacy protection in most cases. However, it is worth noting that Apple MacOS's DP implementation utilizes an epsilon value as high as 6, and Google's DP implementation claims to achieve an epsilon value of 2 in specific scenarios [139]. In our study, we will consider these values as reference points and strive to find a balance between the achieved level of privacy and the system's performance.

5.1.2 Limitations of the State-of-the-Art

The presence of highly sensitive medical information in datasets poses significant challenges to data access. These challenges include concerns for personal privacy and the risk of misuse or re-identification. Data protection laws, such as the EU's GDPR, aim to establish public trust in data sharing and ensure the responsible use of user data by companies. Additionally, the inefficient energy consumption of state-of-the-art forecasting models based on deepNets renders them unsuitable for deployment on edge devices. The inherent resource constraints of edge devices, including limited processing power and memory capacity, pose significant challenges when running complex deepNets models. These models demand substantial computational resources, resulting in a rapid depletion of the limited battery life of edge devices. Therefore, alternative approaches that address the energy efficiency issue are imperative to enable the effective utilization of predictive analytics on edge devices, taking into account their resource limitations.

5.1.3 Proposed Approaches

To address the limited data access issues we propose to generate realistic synthetic data using Generative Adversarial Networks (GANs). Realistic synthetic datasets offer various advantages without compromising user privacy or exposing sensitive information. Firstly, they enhance user privacy by minimizing the risk of re-identification. Additionally, they reduce the likelihood of privacy-breaching attacks on machine learning models, such as the "model inversion" technique [140], [141]. Secondly, these datasets eliminate data that could potentially reveal competitive advantages for the data providers. Despite these modifications, they still maintain fidelity to real-world data. Consequently, the use of realistic synthetically generated datasets has the potential to expedite advancements in machine learning technology. Unlike real data, these synthetic datasets overcome the limitations of availability and can be widely shared and utilized by both industry professionals and researchers without concerns regarding privacy [142–147].

To overcome the energy inefficiency associated with deepNets, we employ SNNs [95]. Unlike deepNets, SNNs are considerably more energy efficient due to the use of spike timing information, which includes latencies and spike rates. Furthermore, information transmission is sparse, as communication only occurs when the membrane potential of a neuron reaches a specific threshold. Additionally, communication in SNNs is a 1-bit activity, which greatly decreases the amount of data transmitted between nodes. Moreover, SNNs substitute MAC operations with adders, further increasing their energy efficiency.

While SNNs have been demonstrated to be energy efficient [96–99], their training can pose challenges due to the non-differentiable nature of the transfer function, which hinders traditional backpropagation. Thus, an appropriate learning mechanism is required for training SNNs. One commonly adopted approach for deep SNNs is to initially train the network with differentiable approximated activation functions using conventional backpropagation. During the testing phase, these functions are substituted with spiking activations to transform the network into an SNN. Our work follows a similar methodology.

The contributions of this chapter are as follows:

- In Sect. 5.4 we proposed a synthetic data generation method for Fitbit dataset with the following contributions:
  - We gather and curate a real-world smart healthcare dataset from users located in various geographical locations.
  - We enhance the acquired smart healthcare dataset by incorporating diverse nutritional and activity patterns that account for factors such as age, ethnicity, geolocation, dietary preferences, and other relevant variables.
  - We present a novel approach that utilizes GANs to generate synthetic time-series data with both categorical and numerical values in a tabular format. Furthermore, we introduce techniques to generate privacy-preserving versions of the synthesized data samples.
  - We have developed realistic synthetic smart healthcare datasets that preserve privacy and provide detailed user profiles, including precise nutritional and activity information. These datasets are made openly available for research purposes.
- In Sect. 5.5 we proposed a SNN model for privacy-preserving time series forecasting with the following contributions:
  - We have developed an innovative system for time-series forecasting of user health data streams using SNNs, which outperforms the existing LSTM-based approach in terms of energy efficiency.

- We have implemented and integrated the proposed forecasting model into a comprehensive end-to-end pipeline for health data streams forecasting, utilizing the clustered FL approach.
- We have integrated ε-differential privacy into our pipeline and thoroughly analyzed its impact on the accuracy of both the baseline and clustered models.
- We conducted a comprehensive comparison between the SNN model and its LSTM-based counterpart using various evaluation metrics such as model size, number of trained parameters, model accuracy, training time, and estimated energy consumption. The evaluation was performed on an augmented real-world Fitbit dataset.

5.2 Hardware

For this study, we utilized Fitbit Charge 2 HR devices. Fitbit wristwatches, also referred to as Fitbit smartwatches, are a type of wearable fitness tracker that offer more advanced features compared to traditional Fitbit bands. In addition to monitoring physical activity, these devices offer capabilities such as music playback, call and text notifications, and integration with mobile apps. Fitbit wristwatches include a built-in GPS for tracking outdoor workouts, a heart rate monitor, and various sensors for monitoring different types of exercises. Furthermore, they are designed to be water-resistant, enabling them to be used for swimming and other water-based activities. With a Fitbit wristwatch, individuals can track their fitness progress, set goals, and monitor their overall health and wellness (Fig. 5.5).

Figure 5.5 Fitbit Charge 2 HR wristwatch tracker [148]

5.3 Dataset

For this study, we used the fitness trackers to observe 25 subjects located in Sweden and Belgium. Data was collected using 12 devices, including 2 consistent participants (one female and one male) and 10 users in circulation. Participants were instructed to document their observations for a duration of at least 60 days, resulting in the collection of over 17 million in measurements related to meal logs, heart rate (HR), calories burned, steps taken, activity, and sleep. To address any gaps in nutritional breakdown data for meals, the researchers leveraged the Nutritionix API. [149] for imputation purposes. The collected measurements were consolidated into three daily entries for each user, encompassing details on the nutritional composition of a meal (breakfast, lunch, or dinner), calories expended during the mealtime, resting heart rate (HR) from the preceding day, and activity records for that specific day. Table 5.1 provides a summary of the data ranges contained in the Fitbit dataset.

Table 5.1 Recorded features of the Fitbit dataset

	Features	Unit	Granularity Recorded	Granularity Aggregated	Range
Macro-nutrients	Fat	gm	food	meal	0.03–25
	Carbs	gm	food	meal	0.01–105
	Protein	gm	food	meal	0.04–55
Calories	Burned	kcal	food	meal	416–1435
Heart rate	Resting	bpm	7.5 s	day	49.5–83.4
Activity	Light	mins	day	day	2–481
	Moderate	mins	day	day	0–211
	High	mins	day	day	0–253
	Sedentary	mins	day	day	600–998

The Fitbit platform collects a substantial amount of individually identifiable data, which necessitates obtaining fully informed consent for data disclosure, leading to a limited number of participants. The main advantage of utilizing this dataset is the availability of private information for each user. However, all individually identifiable data was removed prior to processing using data anonymization.

5.4 Synthetic and Private Smart Health Care Data Generation Using GANs

In this section, we propose an innovative approach that combines a Generative Adversarial Network (GANs) with differential privacy mechanisms to create realistic and private smart healthcare datasets. The proposed approach not only generates synthetic data samples that closely resemble the original dataset but also ensures differential privacy across different scenarios, including learning from a noisy distribution or introducing noise to the learned distribution. To evaluate the effectiveness of our approach, we employ a real-world Fitbit dataset. The evaluation findings indicate that our proposed methodology effectively produces synthetic datasets of excellent quality, maintaining the statistical characteristics of the original dataset while upholding the principles of differential privacy. The content presented in this section aligns with our publication P14.

5.4.1 Data Processing Pipeline

Figure 5.6 illustrates the data processing steps involved in our proposed approach. The following subsections provide details of the steps involved.

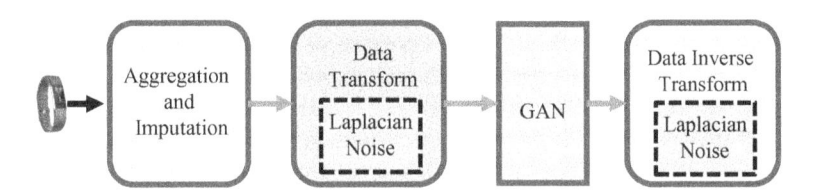

Figure 5.6 The proposed system's data processing pipeline involves aggregating and correcting time-series data, followed by transforming the data to ensure privacy and utility. This includes removing nonessential information, normalizing features, and adding controlled noise for privacy. The model is trained using a novel BGAN architecture to generate synthetic data samples, considering different privacy settings. Finally, the data is inverse-transformed, restoring original ranges, adding missing information, and preserving privacy through noise addition based on the chosen privacy level

5.4.1.1 Data Collection and Imputation

To conduct our experiments for the evaluation of our proposed system, we utilized the dataset provided in Sect. 5.3.

Time-series data aggregation and imputation: Given the manual input nature of the time-series data collection, inherent gaps emerge due to various factors. These gaps result from situations where users fail to wear the device, log meals inconsistently, or when certain meals lack nutritional data. Addressing these data voids involves excluding days without meal logs and meticulously validating remaining entries. When there's no activity profile or discrepancies between recorded burned calories and the profile, we analyze the user's behavior to find similar profiles or refer to prior burned calorie data. A similar strategy is used to reconcile recorded resting heart rate (RHR) and step count against activity profiles.

Meal logs imputation: Filling in missing nutritional breakdowns for meals is more intricate than for other attributes. Among the Fitbit food databases, the US database boasts the most entries, but is tailored to the US region. In contrast, the Belgian (French) database is partially populated. Swedish users, lacking a dedicated database, either selected the closest match in the US database or manually entered custom meal data. Non-English logs were translated to English through a translation API. Entries were replaced with the nearest US food database match or external nutrition API data. The Nutritionix API [149] was used to complete nutritional information. Initially, measurements were grouped into breakfast, lunch, and dinner records, each with details like calories burned, resting heart rate (RHR), step count, and activity history. These were merged into daily records encompassing all meal breakdowns, activity profiles, step counts, calories burned, and RHR. Participants varied behaviors resulted in diverse data, as seen in Table 5.2.

Data Transformation: For data preparation, we start by excluding *Date* and *Gender* information. The remaining features are normalized and used for model training. Depending on the privacy configuration (noisy input), we apply differential privacy (DP) to input data for the GAN, generating DP-synthetic samples per the post-processing theorem. To enhance privacy for categorical or static attributes, we introduce Laplacian noise with $\varepsilon = 0.2$. For behavioral attributes, which pose a lower re-identification risk, Laplacian noise with $\varepsilon = 0.5$ maintains privacy while preserving data utility.

Model Training: As mentioned earlier in Sect. 2.2.5.2, the synthetic data samples are generated using the BGAN (Boundary-seeking GAN) approach. The training of the BGAN model involved selecting the population based on gender and geographical location. Furthermore, the model was trained using three different privacy settings: non-differentially private (non-DP), noisy input, and noisy output. A more detailed explanation of these privacy settings will be provided in Sect. 5.4.3.1.

5.4 Synthetic and Private Smart Health Care Data Generation Using GANs

Table 5.2 Dataset features with ranges (aggregated per day)

Features	Type	Unit	Range
Age	static	yrs	median: 28
Gender	static	–	0: male, 1: female
Height	static	cms	*private*
Weight	static	kgs	*private*
Fat	behavioural	gm	0.08–90
Fiber	behavioural	gm	0.06–34
Carbs	behavioural	gm	0.06–150
Sodium	behavioural	mg	1.92–2745
Protein	behavioural	gm	0.14–75
Calories_burned	behavioural	kcal	1025–4331
Resting_heart_rate	behavioural	bpm	49–83
Lightly_active_minutes	behavioural	mins	2–481
Moderately_active_minutes	behavioural	mins	0–211
Very_active_minutes	behavioural	mins	0–253
Sedentary_minutes	behavioural	mins	254–999
Steps	behavioural	–	162–32871

Data Inverse-transformation: After training and generating data with the generator, we de-normalize features to restore the original data range more accurately. The generated data is then supplemented with *Date* and *Gender* columns to complete the record. We introduce Laplacian noise based on the chosen privacy configuration (noisy output), using $\varepsilon = 0.2$ for static variables and $\varepsilon = 0.5$ for behavioral variables. This ensures privacy while retaining data utility.

5.4.2 Architecture & Learning

5.4.2.1 Generator Network
The architecture of the generator network is shown in Fig. 5.7(a). It takes an input signal of size 15×1, and then passes it through two dense layers with 64 and 32 neurons, respectively. A Leaky ReLU activation function with a rate of 0.2 is applied after each of these layers. Finally, the last dense layer is the output layer, which uses the *tanh* activation function.

(a) Generator Network

(b) Discriminator Network

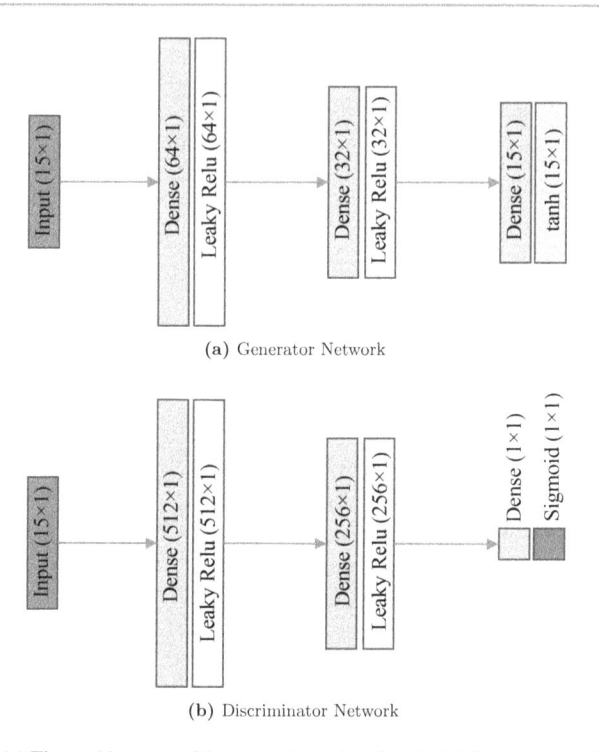

Figure 5.7 (a) The architecture of the generator network, which takes an input signal of size 15 × 1 and passes it through two dense layers with 64 and 32 neurons, respectively. A Leaky ReLU activation function with a rate of 0.2 is applied after each layer, and the final dense layer is the output layer, which uses the *tanh* activation function. (b) The architecture of the discriminator network, which receives a 15 × 1 signal as input and consists of two dense layers with 512 and 256 neurons, respectively. Both layers use Leaky ReLU activation with a rate of 0.2, and the last layer is a dense layer with a single output that applies the *sigmoid* activation function

5.4.2.2 Discriminator Network

The architecture of the discriminator network can be seen in Fig. 5.7(b). The network receives an input of a 15 × 1 signal and consists of two dense layers with 512 and 256 neurons respectively. Both layers use Leaky ReLU activation with a rate of 0.2. The last layer of the discriminator is a dense layer with a single output, which applies the *sigmoid* activation function.

5.4.2.3 Learning Rule
In both the Discriminator Network and the final GAN network, we utilize the adaptive moment estimation (Adam) optimizer. The Adam optimizer dynamically adjusts the learning rate for each weight in the network by estimating the first and second moments of the gradients. The configuration parameters for the optimizer are defined as follows: the learning rate α is assigned a value of 0.0002, while the exponential decay rates for the first moment estimate, β_1, and the second moment estimate, β_2, are set to 0.5 and 0.999, respectively. To address the issue of potential division by zero, we set the epsilon value to $1e^{-8}$.

5.4.3 Results & Discussion

5.4.3.1 Experiments
The proposed pipeline, depicted in Fig. 5.6, incorporates various stages where Laplacian noise is introduced to guarantee differential privacy. This approach leads to three distinct experimental settings. In all scenarios, the GAN network demonstrates successful learning of the data distribution, enabling the generation of realistic samples.
Synthetic Data Generation with no Noise Addition: In this configuration, the GAN network takes the original data as input with the objective of generating results that are close to the original data without differential privacy.
Synthetic Differentially Private Data Generation: Within this configuration, the GAN network undergoes training using differentially privatized data, specifically where differential privacy is applied prior to the data being transmitted to the discriminator (noisy input). This approach allows the GAN model to generate differentially private synthetic samples, which proves beneficial in scenarios involving the outsourcing of synthetic data generation to third-party entities or when the server node itself is included in the threat model.
Applying Differential Privacy to the Synthetic Data: In this configuration, we introduce differential privacy to the generated synthetic data in order to assess the effects of noise addition on the quality of the generated data (noisy output). This approach provides greater flexibility in regulating the amount of noise introduced to the generated data, taking into account the sensitivity of the data features. Nonetheless, this process necessitates additional computational effort to incorporate noise into each individual data point generated.

Our proposed method utilizes BGAN in combination with DP to generate synthetic and private smart healthcare data. The GAN network is robust and capable of producing plausible results. Figure 5.8 illustrates the stability of our GAN model. The upper subplot displays line plots representing the discriminator loss for real

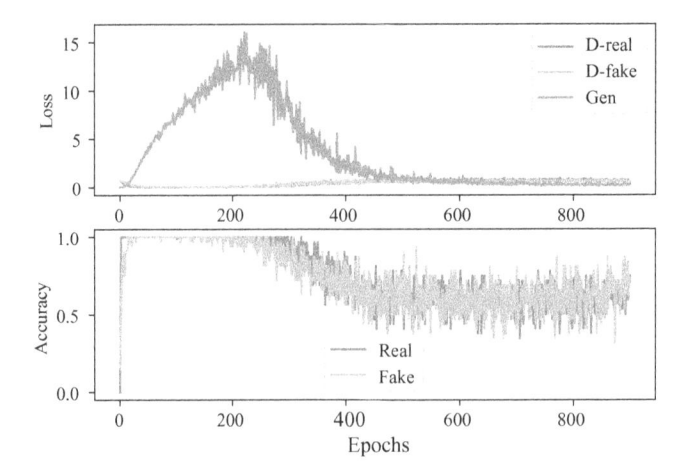

178　　　　　　　　　　　　5 Time Series Forecasting of Healthcare Data

Figure 5.8 The stability of the GAN model is shown. The top subplot illustrates the discriminator loss for real samples (blue), the discriminator loss for generated fake samples (orange), and the generator loss for generated fake samples (green). The losses exhibit instability early in the training phase but become stable between epochs 420 to 600, with slightly increased variance. The discriminator loss for both real and fake samples is around 0.5, while the generator loss ranges between 0.5 and 1.0, indicating that the model is expected to produce plausible samples during this period

samples (blue), the discriminator loss for generated fake samples (orange), and the generator loss for generated fake samples (green). The observations reveal that these three losses experience some instability during the initial stages of training, but reach stability between epoch 420 and epoch 600. Subsequently, the losses remain relatively stable, albeit with a slightly increased variance. The discriminator loss, both for real and fake samples, tends to hover around 0.5, while the generator loss fluctuates slightly higher, ranging from 0.5 to 1.0. Consequently, the model is anticipated to generate plausible samples within the epoch range of 420 to 600.

Table 5.3 presents a subset of rows from the original dataset alongside synthetic rows generated by the trained GAN models. The datasets are organized into different categories based on the type of privacy settings applied. Specifically, the categories include *Original* and *GAN*, representing non-differentially private datasets without noise addition. The category *GAN with DP output* showcases synthetic data samples generated with the addition of differential privacy noise, indicating a noisy output. Similarly, the categories *Original DP* and *GAN with DP input* represent the original differentially private input and the corresponding synthetic samples, respectively,

5.4 Synthetic and Private Smart Health Care Data Generation Using GANs

Table 5.3 Example data samples from the population in Belgium with hidden attributes of age and gender

Dataset	Height	Weight	Fat	Fiber	Carbs	Sodium	Protein	Calories Burned	Resting HR	Lightly	Moderately	Very	Sedentary	Steps
Original	169	66.18	39.0	11.0	33.0	1189.0	4.0	2308.08	59.526	121	6	28	731	9706
	169	66.18	39.0	7.0	125.0	1125.0	34.0	2707.35	61.99	166	13	56	732	14070
	169	66.18	33.0	8.0	96.0	1361.0	66.0	2485.35	58.45	99	22	60	774	12008
BGAN	175	87.09	29.20	7.80	73.46	1192.83	41.48	2556.28	63.44	157	63	78	768	12222
	175	87.09	41.32	10.71	49.00	1072.10	33.07	2873.35	64.92	195	49	84	772	14374
	175	87.09	60.82	11.69	98.62	1447.21	31.67	3286.82	64.77	253	56	73	889	14877
BGAN w/ DP output	177	93.62	30.51	13.55	70.19	1191.99	42.48	2555.69	66.29	154	60	71	772	12222
	177	93.62	37.94	9.63	50.15	1071.65	34.7	2875.86	70.71	195	48	81	772	14374
	177	93.62	62.06	9.84	94.18	1447.51	32.17	3286.16	70.32	252	49	73	892	14875
Original DP	161	66.78	39.04	10.02	33.97	1195.83	5.42	2308.34	57.32	121	8	26	732	9704
	161	66.78	38.06	7.31	125.42	1122.46	32.66	2706.42	67.77	164	9	56	732	14074
	161	66.78	29.79	2.82	99.02	1360.55	65.02	2485.37	57.75	97	23	62	777	12005
BGAN w/ DP input	181	80.74	33.14	4.12	99.12	1020.8	39.5	3099.14	58.69	213	51	23	757	9769
	181	80.74	22.25	11.02	38.29	1416.57	4.838	2611.83	58.09	137	2	3	754	9944
	181	80.74	58.99	11.19	104.60	483.94	41.96	2593.24	59.56	190	48	106	732	12004

indicating a noisy input. Notably, all the generated examples demonstrate a realistic appearance across all selected privacy settings.

In order to evaluate the similarity between the generated data and the actual data, we visually represent their respective histograms. Specifically, our focus lies on the distribution of burned calories among male participants in Belgium. Figure 5.9 (a) and (b) display the histograms for the original and synthesized burned calories, respectively. The similarity observed between the two distributions signifies the efficacy of our proposed BGAN network. Additionally, we employ the Kolmogorov-Smirnov (KS) goodness-of-fit test [150] on the samples extracted from the original and synthesized distributions. The test yields a high *p-value* of 0.98, indicating a strong likelihood that the samples originate from the same distribution. This outcome demonstrates the ability of our BGAN model to learn various categorical and numerical features and generate realistic synthetic data.

Figure 5.9(d) showcases the distribution of the original DP data samples with noisy input, which bears a close resemblance to the distribution of the DP data generated by the BGAN, as depicted in Fig. 5.9(e). Moreover, the KS test conducted on the distribution of the original noisy calories (DP input) and the synthetically generated noisy calories distribution yields a high *p-value* of 0.97. This high *p-value* suggests a strong probability that the samples originate from the same distribution. These outcomes serve as confirmation that the proposed BGAN possesses the ability to generate distributions of differentially private samples.

By utilizing DP input, we are able to generate differentially private data without the need to apply DP to all the synthesized data samples explicitly. Conversely, applying the DP mechanism post-synthetic data generation offers greater control in terms of the level of noise addition, resulting in better data utility. The distribution of the samples, as illustrated in Fig. 5.9(c), is preserved despite the records being noised and differentially private.

Our experiments and results confirm that despite the simplicity of the proposed GAN architecture, it achieves exceptional performance in generating both synthetic and differentially private synthetic data. The smart healthcare dataset based on Fitbit contains highly diverse features, and the DP mechanism with BGAN proposed in this study is both stable and produces synthetic data with high utility.

5.4 Synthetic and Private Smart Health Care Data Generation Using GANs 181

(a) original calories

(b) synthetic calories

(c) Synthetic noisy output calories

(d) Original noisy calories

(e) Synthetic noisy calories

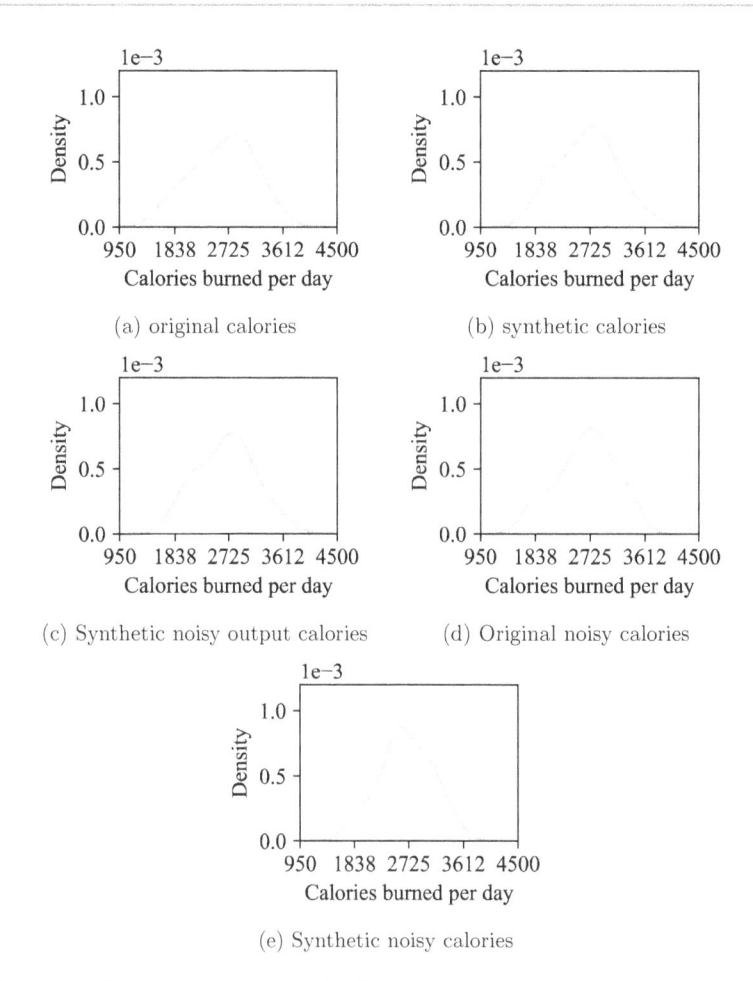

Figure 5.9 The histograms illustrate the distribution of calories burned per day (kcal) for a specific group: Belgium males with a resting heart rate (RHR) of RHR=70-75bpm. Figure (d) represents the distribution of the original differentially private (DP) data samples with noisy input, while Figure (e) displays the distribution of the DP data generated by the BGAN. The resemblance between the two distributions is supported by a high p-value of 0.97 obtained from the Kolmogorov-Smirnov (KS) test conducted on the original noisy calorie distribution and the synthetically generated noisy calorie distribution. These findings validate the effectiveness of the proposed BGAN model in generating distributions of differentially private samples

5.4.4 Conclusion

We presented a system for generating synthetic and private smart healthcare datasets using BGANs and differential privacy. Our approach was evaluated on a collection of real-world Fitbit-based smart healthcare datasets under three privacy preservation settings. Our method demonstrated the ability to effectively learn and generate diverse categorical and numerical values, resulting in stable GANs for dataset generation. The synthetic datasets produced were realistic and exhibited similar distributions as the real data while maintaining user privacy. Our method also provides flexibility in controlling different privacy settings, allowing for the publication of open smart healthcare datasets that can be used for research and industry applications.

5.5 Time-Series Forecasting on User Health Data Streams

In this section, we investigate the utilization of SNNs for time-series forecasting, capitalizing on their well-established energy-saving capabilities. Our research primarily revolves around developing an energy-efficient and privacy-preserving forecasting system using SNNs, specifically applied to real-world health data streams. To provide a comparative analysis, we benchmark our system against a state-of-the-art approach employing LSTM-based prediction models. The evaluation results demonstrate that although SNNs introduce an accuracy trade-off (with a 2.2× greater error), they offer a more compact model size (19% fewer parameters) and significantly reduced memory consumption (77% less). Moreover, SNNs exhibit a 43% reduction in training time and an estimated energy consumption of 3.36 µJ, which is notably lower than that of traditional ANNs. In our pursuit of enhanced privacy protection, we integrate ε-differential privacy into our federated learning-based models, ensuring robust privacy guarantees. By incorporating a differential privacy parameter of $\varepsilon = 0.1$, our experiments reveal only a minimal increase in the measured average error (RMSE) of 25%. The content presented in this section aligns with our publication P9.

5.5.1 Data Processing Pipeline

In this section, we present the data processing pipeline employed in our health data prediction system. The data streams extracted from Fitbit devices encompass various aspects, including the user's daily meal logs, indicating their consumption

5.5 Time-Series Forecasting on User Health Data Streams

of carbohydrates, fats, and proteins, as well as information on calories burned, resting heart rate, and active minutes. Our forecasting system consists of two key components: (1) a clustering mechanism utilizing streaming k-means and pattern matching to group users into k clusters based on their meal logs, and (2) a health prediction system that employs FL-based models for each user group to forecast their health and meal data. To ensure privacy preservation, we integrate DP into our system. Figure 5.10 provides an overview of the health forecasting system pipeline, illustrating the clustering and prediction components with DP integration.

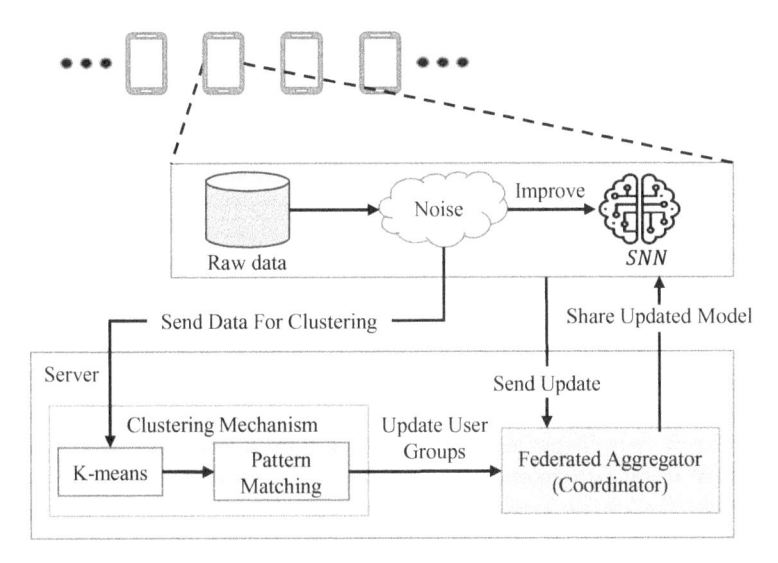

Figure 5.10 The health forecasting system's pipeline involves clustering and prediction components with differential privacy (DP). The clustering uses streaming k-means to group users with similar eating patterns and pattern matching to find similarities. FL is employed to train distinct models for user clusters from clustering, adding Laplacian noise via a noise addition mechanism for DP. Predictions are made locally on clean data

5.5.1.1 The Clustering Mechanism

The health data collected from users' daily meal logs is processed using the streaming k-means algorithm to cluster users with similar eating patterns. The meal logs encompass breakfast, lunch, and dinner data. The initial centroids are selected as the first unique meals, and subsequent meals are assigned to these centroids based on their Euclidean distance. Following the clustering of logs over a span of seven days, pattern matching is performed to group users with similar eating habits. Each user's pattern is generated using centroid IDs. For example, in Fig. 5.11, the blue bars represent the centroid IDs for three meals (breakfast, lunch, and dinner) across three users. These clusters enable the determination of each user's eating pattern: user1 (0,0,0), user2 (0,0,1), and user3 (1,1,1). The Hamming distance [151] is calculated to evaluate the similarities between eating patterns, allowing users with the lowest Hamming distance to be placed in the same user group. Finally, FL is employed to train the forecasting model per group, as well as a central generic model.

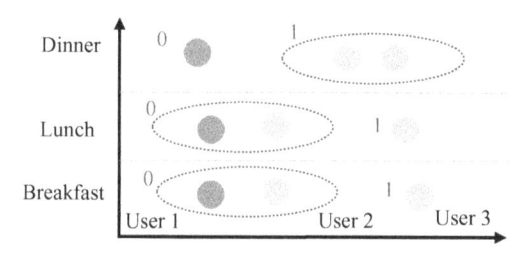

Figure 5.11 The clustering process groups users with similar eating patterns based on centroid IDs for breakfast, lunch, and dinner from three users. Eating patterns are defined by Hamming distance from these clusters. Users with the closest Hamming distance are grouped. The forecasting model is trained per group using Federated Learning (FL) and a central generic model

5.5.1.2 Data Model for Training SNNs

The dataset consists of time-series health data streams, encompassing meal logs, active minutes, resting heart rate, and calories burned. We leverage this dataset to train our forecasting models, aiming to make one-step predictions or estimate a user's health and meals based on their historical data. To accomplish this, we employ the multiple-input single-output (MISO) methodology, also referred to as single-step prediction. To forecast the next element x_{n+1} in a given time-series $[x_1, x_2, x_3, ..., x_n]$ using an input array of size M, the multiple-input single-output (MISO) approach entails learning a function $\hat{F}$ defined as follows:

$$\hat{F}(x_{t+1}, ..., x_{t+M}) = \hat{x}_{t+M+1}, \tag{5.1}$$

where $\hat{F}$ takes as input the previous M data points $x_{t+1}, ..., x_{t+M}$ and generates a single-step estimation of the next data point, denoted as $\hat{x}_{t+M+1}$. This sliding window prediction mechanism employs a window length of M, which is a design choice, and the window is incremented by 1 for each prediction step.

5.5.1.3 Clustered Federated Learning and Differential Privacy

In the clustering-based FL method, separate FL models are created for user clusters identified by the clustering approach. For DP, Laplacian noise is added to the data using a noise mechanism. During learning, each participant model trains on noise-introduced data to enhance the FL model or its corresponding model in the clustered FL system. Figure 5.10 shows the noise addition process used for FL in clustering. As clustering works with noise-affected data for training user clusters, cluster composition may vary in different trials. It's important to highlight that predictions are made locally on noise-free data.

5.5.2 Architecture & Learning

In our novel SNN architecture for time-series forecasting, we proposed employing Legendre Memory Units (LMUs) [152] as illustrated in Fig. 5.12. The NengoDL simulator [36] is used to construct the network, which offers a spiking version

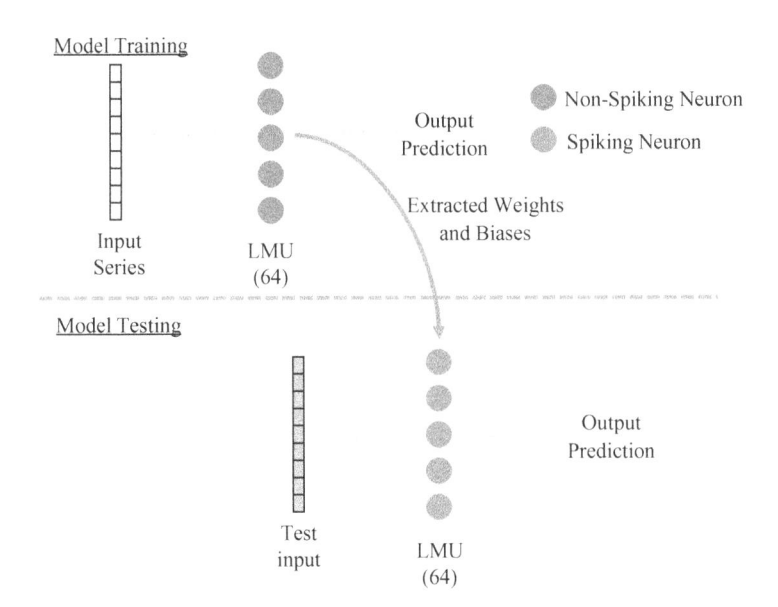

Figure 5.12 The proposed SNN architecture involves training the model with non-spiking neurons using backpropagation. Subsequently, the trained model is reconstructed with spiking neurons, while utilizing the weights and biases obtained from the pre-trained model

of rate-based neurons. The training of the network is accomplished through conventional backpropagation with a non-spiking *Tanh* activation function. An LMU network involves 3 core parameters: units, order, and theta. In our case, we utilize 64 neurons as units. The order pertains to the LMU basis functions and represents the number of Legendre polynomials utilized for the orthogonal representation of the sliding window. Our configuration adopts an order of 128. Increasing the LMU order enhances the representation and retention of swift alterations in the input signal. On the other hand, theta refers to the duration of time that is retained within the internal memory of the LMU. This parameter influences the length of time that the basis functions represent, and subsequently, the amount of temporal information preserved within the LMU network for later utilization in training and decoding. In our approach, we set theta to 360.

5.5.2.1 Loss Function

The loss function we use to evaluate the accuracy of our forecasting is the Root Mean Square Error (RMSE). We assign greater importance to large errors since our objective is to attain precise predictions. Mathematically, RMSE is represented as:

$$RMSE = \sqrt{\frac{1}{n}\sum_{i=1}^{n}(Y_i - \hat{Y}_i)^2}, \quad (5.2)$$

where Y_i represents the actual values, $\hat{Y}_i$ represents the predicted values, and n denotes the total number of samples.

5.5.2.2 Learning Schedule

During the learning process, we employ the adaptive moment estimation (Adam) technique for stochastic gradient descent, which dynamically adjusts the learning rate based on first and second moment estimates. Specifically, we set the learning rate to 0.003, while the exponential decay rates for the first moment (β_1) and second moment (β_2) estimates are assigned values of 0.9 and 0.999 respectively. To maintain numerical stability, we set epsilon to $1e^{-7}$.

5.5.2.3 Model Testing

During the testing phase, we employ spiking *Tanh* neurons to reconstruct the trained network and establish a SNN. The weights of the trained model are extracted and utilized for the spiking *Tanh* neurons. To obtain precise measurements of the spiking

5.5 Time-Series Forecasting on User Health Data Streams

neuron output over time, we repeat the test inputs multiple times and feed them into the network.

5.5.2.4 Model Hyperparameters

The study involves hyperparameters derived from three distinct categories: the spiking neuron model, LMU, and the model itself. These hyperparameters are determined via empirical methods and are documented in Table 5.4.

Table 5.4 The hyperparameters used for training the proposed SNN

Hyperparameter	Value	Definition
Type of Neuron	Spiking Tanh	Neuron model used
τ_{ref}	0.005	Absolute refractory period expressed in seconds
LMU's units	64	Number of employed neurons
LMU's order	128	Number of Legendre polynomials
LMU's theta	360	Amount of time kept in the LMU's internal memory
Learning	0.003	Learning rate of our Adam optimiser
Minibatch Size	100	Number of concurrent inputs fed to the network
Epochs	10	Number of passes of the entire training dataset
Delay	10 ms	Latency of the output in respect to the input
Inference's time steps	50	Number of inferences of SNN before results

5.5.3 Evaluation Methodology

During our experimental analysis, we assess the accuracy, model size, and number of network parameters associated with our SNN model and conduct a comparative analysis with the results obtained using LSTM. We evaluate these models using both clustered and non-clustered user data. Furthermore, we investigate the effect of privacy guarantees on prediction accuracy by comparing the accuracy achieved with and without the integration of privacy-preserving techniques.

5.5.3.1 Dataset

For this experiment, we used the dataset described in Sect. 5.3 and the augmented dataset obtained with the proposed GAN model in Sect. 5.4.

Data Augmentation Using Generative Adversarial Networks: We utilize the augmented Fitbit dataset as a practical example of a health and nutrition monitoring dataset, which offers real-world insights. Due to the inclusion of sensitive information within the dataset, it serves as a valuable resource for evaluating the effects of implementing privacy-preserving methods on prediction accuracy.

The integration of the datasets yielded a dataset comprising 630 users, where each individual had logs spanning approximately two months on average. Table 5.5 provides a concise overview of the Augmented Fitbit Dataset, including the features, their corresponding units, and the ranges of values they encompass. The feature labeled as *Macronutrients* comprises the *carbohydrate*, *protein*, and *fat* content of the meals recorded at each timestamp. The *Calories burned* at time t represents the number of calories expended by the user between time $t-1$ and time t, while the *Active minutes* follow the same principle. The *Resting heart rate* is calculated over the entire day.

Table 5.5 Value range of the Augmented Fitbit dataset

Features	Unit	Min. Value	Max. Value
Macronutrients	grams	0.313	1463.108
Calories burned	kCal	467.006	2335.445
Resting heart rate	bpm	55.029	89.995
Active minutes	minutes	0	1242.413

5.5.3.2 Performance Metrics

In the learning phase, we assess the accuracy of our model using RMSE, which is illustrated in Equation 5.2. To explicate the performance evaluations for the FL models, we assume that m is the total number of clients participating in the training process. We denote the local datasets for each client as $X_1, X_2, ..., X_m$. These datasets are divided as follows: 80% of the data is used for training, and 20% is used for testing for each local dataset. To determine the overall performance, we calculate and average the selected errors from each local test dataset using the following equation:

$$RMSE_{federated} = \frac{1}{m}\sum_{i=1}^{m} RMSE_i. \tag{5.3}$$

5.5.4 Results & Discussion

In this section, we present the outcomes of our experiments conducted on the augmented Fitbit dataset. The experiments encompass four configurations: 1) the baseline model trained on the complete dataset, 2) the FL model, 3) the differential private baseline model, and 4) the differential private FL model. Furthermore, we thoroughly analyze and discuss the results obtained from these experiments.

5.5.4.1 Baseline Model

The accuracy, training time, and model size of the baseline model are assessed using both SNNs and LSTMs.

Prediction Accuracy: To evaluate the accuracy of our predictions, we adopt the single-step prediction approach, wherein we forecast the features for the subsequent meal (Breakfast, Lunch, or Dinner). The accuracy achieved by both the SNN and LSTM models is documented in Table 5.6. The results indicate an increase in error, with an average error approximately 2.17 times higher than that attained by the LSTM model. Although the SNN model does not match the predictive performance of LSTM in capturing users' health habits, it still demonstrates a relatively high level of accuracy within the range of values presented in Table 5.5. In fact, our forecasts deviate by only 0.331% from the ground truth values for each feature range.

Table 5.6 Comparison of prediction accuracy between the SNN model and the LSTM baseline model, using the units specified in Table 5.5

Predicted	RMSE LSTM [153]	RMSE SNN	Change in error
Macronutrients	1.540	3.364	+2.19x
Calories burned	14.460	31.732	+1.19x
Resting heart rate	0.0058	0.298	+4.13x
Active Minutes	2.983	6.457	+1.16x

Model Size: As illustrated in Table 5.7, our proposed SNN model exhibits a notable reduction in the number of parameters, with approximately 19% fewer parameters compared to the LSTM model. Moreover, the SNN model is 77% lighter in terms of its overall size when compared to the LSTM model.

Table 5.7 Comparison of the number of parameters and model size (in bytes) between the SNN and LSTM models

Metric	LSTM	SNN	Change
# Parameters	36003	28,929	−19.6%
Memory size (bytes)	483784	108856	−77.5%

5.5.4.2 Clustered Federated Learning

In this experiment, we trained single-feature models for each feature within every cluster. Table 5.8 showcases a significant improvement in accuracy for two of the four clusters compared to our baseline performance based on the SNN model. Fur-

Table 5.8 Comparison of accuracy in clustered FL models utilizing SNNs. The table illustrates the change in error relative to the baseline model trained on the entire dataset

Clusters	Total users	Predicted	RMSE	Change in error	Average training time
Cluster 1	209	Macronutrients	1.983	−41.1%	434s
		Calories Burned	18.071	−43.2%	
		Resting Heart Rate	0.166	−44.3%	
		Active Minutes	3.778	−41.5%	
Cluster 2	171	Macronutrients	2.353	−30.1%	387s
		Calories Burned	25.985	−18.2%	
		Resting Heart Rate	0.202	−32.2%	
		Active Minutes	4.751	−26.4%	
Cluster 3	56	Macronutrients	3.728	+10.8%	204s
		Calories Burned	37.742	+19%	
		Resting Heart Rate	0.375	+25.8%	
		Active Minutes	7.533	+16.6%	
Cluster 4	33	Macronutrients	5.192	+54.3%	147s
		Calories Burned	58.184	+83.4%	
		Resting Heart Rate	0.471	+58.1%	
		Active Minutes	10.464	+62%	

thermore, Cluster 1, despite having a longer training duration (with a higher number of seconds), exhibits a training speed that is over 30 times faster than our baseline model. However, our baseline model outperforms Clusters 3 and 4. We can deduce that the impact of cluster size on prediction accuracy only arises when the model lacks sufficient data for learning. In cases where a cluster contains adequate data, leveraging the similarity among its components contributes to improved results. Consequently, the similarity among users can be considered as a characteristic that triggers enhanced prediction accuracy.

5.5.4.3 Differentially Private Learning

We explore the implications of incorporating privacy-preserving techniques into the system, particularly within the baseline model. Our approach aims to ensure privacy by introducing noise into the data. To obtain reliable results, we conduct the experiments five times and compute the average outcomes for each configuration. Table 5.9 depicts the decline in accuracy corresponding to escalating levels of privacy implemented during the training data phase. Google [139] has demonstrated the achievability of differential privacy with $\varepsilon = 2$ under specific conditions. Hence, we commence our experiments with $\varepsilon = 2$, incorporating Laplace noise with a mean of 0 and a variance of $\frac{1}{\varepsilon}$. As demonstrated in Table 5.9, we successfully achieve this level of differential privacy with a mere average increase of 2% in prediction

Table 5.9 Evaluation results after introducing noise to the training data to achieve differential privacy (DP) in the baseline model. The table presents the average change in error compared to the baseline model trained on the complete dataset

Predicted	RMSE				
	$\varepsilon = 2$	$\varepsilon = 1$	$\varepsilon = 0.1$	$\varepsilon = 0.025$	$\varepsilon = 0.01$
Macronutrients	3.408	3.507	4.142	4.863	7.533
	(1.30 %)	(4.25%)	(23.13%)	(44.56%)	(123.93%)
Calories Burned	32.541	32.763	39.891	46.724	59.263
	(2.55 %)	(3.25%)	(25.71 %)	(47.25 %)	(86.76 %)
Resting Heart Rate	0.307	0.315	0.387	0.428	0.579
	(3.02 %)	(5.70%)	(29.87 %)	(43.62 %)	(94.30 %)
Active Minutes	6.587	6.694	7.850	9.941	13.071
	(2.01 %)	(3.67%)	(21.57 %)	(53.96 %)	(102.43 %)
Average change in error	2.22%	4.22 %	25.07 %	47.35 %	101.86 %

error. Additionally, our analysis shows that a viable privacy-accuracy balance can be achieved using ε values from 1 to 0.025. Our experiments indicate that the optimal trade-off happens at $\varepsilon = 0.1$, resulting in around a 25% increase in prediction error. This leads us to explore the implications of applying this privacy level ($\varepsilon = 0.1$) within the clustered FL configuration.

5.5.4.4 Differentially Private Federated Learning

We will now assess the impact of adding DP noise on the user clustering in our FL setup. Each participant in the learning process trains on the noised data and attempts to enhance the federated model. It is important to note that, when training on clusters of users, the clustering mechanism also receives noised data, resulting in clusters that are different for each experiment. It is worth noting that in comparison to our second experiment (Sect. 5.5.4.2), as the noise in the data increases, there are fewer users per cluster overall. As a result, the algorithm identifies a larger number of smaller clusters. This occurs because the degree of noise in the data is inversely proportional to the similarity between users.

The accuracy results obtained from clustering users with $\varepsilon = 0.1$ DP-noise addition are presented in Table 5.10. Notably, the table demonstrates that user clustering continues to exert a positive influence on overall accuracy, even in the presence of added noise in the data. Based on our findings, it is evident that despite the introduction of noise to the data, users retain a certain level of similarity that can be harnessed to enhance prediction accuracy. This suggests that even with the implementation of privacy-preserving techniques, the clustering mechanism continues to have a positive impact on overall accuracy. However, our results also indicate that when clusters are small, as observed in the case of Cluster 4 with only 34 users, the model encounters challenges in training effectively due to limited data availability. The outcomes of this experiment exemplify the model's ability to maintain high accuracy while experiencing only a marginal performance decline compared to the non-noised model. It is noteworthy that the addition of a moderate level of noise to the data can serve as a regularization technique, aiding in accuracy improvement and mitigating overfitting. However, excessive noise can have a detrimental effect on the model's performance.

Table 5.10 Comparison of outcomes achieved by incorporating noise to the training data to achieve differential privacy in the clustering configuration with $\varepsilon = 0.1$. The average change in error is presented in relation to the baseline model trained on the complete dataset, employing differentially private learning (Sect. 5.5.4.3). Increases in the average change in error reflect a decrease in accuracy

Clusters	Total users	Predicted	RMSE	Change in error	Average training time
Cluster 1	158	Macronutrients	2.821	−31.89%	376s
		Calories Burned	26.078	−34.63%	
		Resting Heart Rate	0.253	−34.63%	
		Active Minutes	5.272	−32.84%	
Cluster 2	101	Macronutrients	3.058	−26.17%	268s
		Calories Burned	31.853	−20.15%	
		Resting Heart Rate	0.301	−22.22%	
		Active Minutes	6.747	−14.05%	
Cluster 3	71	Macronutrients	3.601	−13.06%	214s
		Calories Burned	33.445	−16.16%	
		Resting Heart Rate	0.344	−11.11%	
		Active Minutes	6.681	−14.89%	
Cluster 4	34	Macronutrients	4.277	+3.26%	151s
		Calories Burned	40.172	+0.70%	
		Resting Heart Rate	0.396	+2.33%	
		Active Minutes	8.034	+2.34%	

5.5.4.5 Energy Estimation of SNN Model

To estimate the energy usage of the proposed model, we utilize the hardware metrics of the (μBrain) chip described in [154]. The energy consumption for each classification inference is defined as:

$$J_C = \eta \times \zeta + \delta T \times \Lambda$$

where η represents the maximum number of spikes during forecasting, which is equal to 3247. ζ represents the energy consumed per spike, which is 2.1 pJ. Λ represents the static leakage power, which is 73 µW . δT represents the inference time, which is 46 ms. Based on these values, we estimate the energy consumption of our model to be $J_C = 3.36\,\mu J$ per inference. Therefore, our SNN is expected to be quite energy efficient when predicting values on spiking neuromorphic hardware as compared to deep neural network solutions [113].

5.5.5 Conclusion

Our study highlights the efficacy of SNNs in time series forecasting in terms of their ability to provide acceptable accuracy while consuming less memory and less training time, making them suitable for edge devices. Furthermore, the development of a clustered FL-based health data prediction model indicates that leveraging user similarities can enhance forecasting model performance, although the success of the clustering method depends on the data. When clusters have insufficient data to train the algorithm, prediction accuracy suffers greatly. Additionally, our results demonstrate that our clustering FL approach benefits from ε-DP, with a notable improvement in prediction accuracy observed when similar users are grouped together.

Conclusion and Future Directions 6

The thesis explores the effectiveness of spiking neural networks (SNNs) for edge devices, with a focus on resource efficiency, a key requirement for edge computing. The study primarily considers consumer radar as an application area, with a specific use case of gesture sensing and air-writing. To generalize the scope of this dissertation, the Fitbit tracker application is also utilized, covering a use case of time series forecasting for health data.

In the context of gesture sensing, the study first evaluates the performance of SNN classification using a traditional signal processing chain, which involves using range-Doppler images for classification. Subsequently, the study demonstrates how to simulate signal processing blocks such as range FFT and Doppler FFT with SNN to avoid extra FFT computation and the requirement of an additional microcontroller to perform these operations. Furthermore, the study also explores classification directly from raw data.

In the realm of air-writing, the study shows the ability to classify and recognize characters using a single radar, as opposed to state-of-the-art methods that require three or more radars. The study proposes eliminating the feature extraction stage and letting the neural networks and SNN learn intrinsically. Moreover, the study proposes not only classifying but also reconstructing the original drawn character. Additionally, the study proposes the use of spiking neural networks for classification and recognition of characters drawn in the air.

Finally, regarding time series forecasting of Fitbit healthcare data, the study proposes the generation of synthetic data using GANs and presents an evaluation of a privacy-preserving solution for healthcare data stream forecasting. In this context, the study takes into account the trade-offs between preserving user privacy, application runtime, and prediction accuracy.

© The Author(s), under exclusive license to Springer Fachmedien Wiesbaden GmbH, part of Springer Nature 2024
M. Arsalan, *Optimization of Spiking Neural Networks for Radar Applications*,
https://doi.org/10.1007/978-3-658-45318-3_6

6.1 Summary of the Results

Chapter 3 outlines 5 novel approaches to gesture sensing utilizing SNNs. The first approach suggests a gesture sensing system that operates on a conventional signal processing pipeline of range-Doppler. This system's results were compared to the state-of-the-art conventional neural networks, and the proposed model achieved similar levels of accuracy for four gestures, specifically 98.5% compared to the LSTM and CNN-LSTM methods, which achieved 96.9% and 97.18% respectively.

The subsequently proposed approach involves a gesture sensing system utilizing spiking neural networks, which excludes the use of convolutional layers to satisfy hardware restrictions. This is accomplished by refining the SNN architecture. The proposed model yields an accuracy of 97.5% for four gestures, which is comparable to the LSTM and CNN-LSTM methods' accuracy levels.

Subsequently, we proposed a SNN model that could classify up to 8 complex gestures. The preprocessing pipeline underwent enhancements to generate more informative features that can be effectively learned by the predictive network. To improve the system's robustness, comprehensive hand profiles, including range-time, velocity-time, and angle-time images, were employed as features instead of range-Doppler images. A modified and enhanced SNN architecture was introduced, achieving a high accuracy of 99.5% in detecting and identifying eight gestures. The proposed model has a compact size of only 75 kB, which is considerably smaller compared to existing leading models, resulting in efficient memory utilization. Moreover, the energy consumption per classification by the proposed system is significantly reduced to 2.04 µJ compared to deepNets. Furthermore, an in-depth evaluation of various spiking neuron models was conducted on an augmented dataset, revealing that the leaky-integrate and fire model yielded the best performance.

Next, we propose end-to-end pipeline for radar-based gesture sensing using SNNs on raw ADC data. Unlike previous approaches that relied on slow-time and fast-time FFTs, our proposed pipeline eliminates the need for these computations by mimicking the signal pre-processing step (slow-time FFT) directly into the SNN architecture. This integration significantly reduces the overall end-to-end latency by more than a factor of two compared to previous methods, while maintaining a comparable accuracy level of 98.1% for four gestures. Furthermore, the energy consumption per classification is minimized to 2.05 µJ. These advancements in the system highlight the potential of SNNs for efficient and effective radar-based gesture sensing.

Afterward, we propose an end-to-end radar-based gesture sensing system that leverages SNNs to perform gesture recognition using raw ADC data. Unlike previous methods that relied on Doppler images, our proposed approach eliminates the

6.1 Summary of the Results

need for pre-processing steps such as slow-time and fast-time FFTs. This reduction in computational overhead eliminates the requirement for additional computational units. Additionally, our novel approach not only mimics the fast-time FFT but also incorporates the slow-time FFT within the SNN. As a result, the system is capable of classifying up to 8 gestures, expanding upon the previous approach that was limited to four gestures. Furthermore, our proposed model has a size of 14824Kb and exhibits an estimated energy consumption per classification of 2.1 µJ. These advancements contribute to a more efficient and versatile gesture sensing system. These system improvements demonstrate the potential for efficient and effective radar-based gesture sensing using SNN. Furthermore, we have shown the quantization effects on the model performance with 4, 8 and 16 bits quantization.

Finally, in Chap. 3, we propose the implementation of an FPGA-based gesture sensing system utilizing SNNs. During the deployment process, we encountered challenges that shed light on the immaturity of state-of-the-art tools for SNNs when it comes to implementing them on hardware prototypes. This observation was based on the substantial amount of time and effort required to transform the Nengo model into a format suitable for efficient execution on both the host and the SNN FPGA platform. Despite these challenges, our implementation on Xilinx FPGAs achieved a remarkable level of performance, with an accuracy of 98.5%, which is comparable to other approaches.

In Chap. 4, we proposed four novel air-writing systems. First, we propose a novel air-writing system based on a sparse network of radars, using a novel 1D DCNN-LSTM-1D transpose DCNN architecture. This system allows for the reconstruction and recognition of drawn characters using only the range information from each radar. Unlike traditional radar-based air-writing systems, this approach does not rely on trilateration algorithms, which can fail if the finger or hand target is not detected by all three radars. Additionally, the proposed system does not require hand-crafted feature images, as the deep neural network learns these features intrinsically. The proposed system exhibits superior performance compared to the baseline methods, achieving an accuracy of $97.33\% \pm 2.67\%$ when two radars are utilized, while the best accuracy achieved by the baseline method is 98.33%. In the case of a single radar, our proposed method attains an accuracy of $90.33\% \pm 4.44\%$. In terms of reconstruction, the average reconstruction MSE for single radar is & $6.8 \exp-4 \pm 1.5 \exp-4$ and for two radars $1.4 \exp-4 \pm 0.7 \exp-4$. We demonstrate the success of the proposed system through real 60-GHz sensor data and show that it is highly accurate, paving the way for easy practical deployment of such systems in low-commodity hardware.

We introduce a new method for air-writing recognition using one or two radars to track hand movement and a 1D temporal convolutional network for simultaneous feature extraction and character recognition. This approach excels not only in character recognition but also in continuous writing tasks, achieving impressive accuracy rates: 99.11% and 91.33% for two radar-based solutions and one radar-based solution, respectively. These results outperform other deep architectures, highlighting the system's superiority in handling continuous writing. Character segmentation within continuous writing sequences is accomplished using thresholding, yielding a segmentation accuracy of 100% across over 50 words with character counts spanning 2 to 4. The models also boast smaller memory footprints, with sizes of 643kB (single radar) and 644kB (two radars). The successful demonstration of air-writing recognition via local trajectory using Infineon's 60-GHz FMCW radar sensor data shows promising results for practical deployment in space or cost-constrained systems.

Next, we propose a highly energy-efficient air-writing system that employs an SNN to classify and recognize the trajectory of characters drawn in the air, making it ideal for edge IoT devices where energy efficiency is a primary concern. The proposed SNN model achieves a classification accuracy of 98.6% and is of relatively small size of $3.7MB$, demonstrating its memory efficiency in terms of storage requirements. The estimated energy per the classification of the proposed model is the energy consumption per the classification of the proposed system is 2.13 µJ. To the best of our knowledge, the SNN architecture proposed for the classification of drawn trajectory with a network of 3 radars using trilateration is the first of his kind and has not been used in the context of air-writing before.

Finally, we present a novel air-writing system based on a single radar using Spiking Legendre Memory Unit (SLMU). We propose using Genetic Algorithm to find the optimal parameters of SLMU for the given application to avoid manual tweaking of the network, which can be very time-consuming. The proposed solution achieves a classification accuracy of 98.53% in the case of two radars and 95.37% in the case of a single radar, with an estimated energy consumption of 2.04 µJ, which is significantly lower than the deep learning counterparts. Moreover, the proposed SNN architecture is memory-efficient, with a storage memory requirement of 490kB for a single radar and 564.2kB for two radars.

In Chap. 5, first, we develop a solution to address the challenges of privacy-preserving data sharing in smart healthcare. Our proposed approach involves designing, implementing, and evaluating a generative adversarial network (GAN) model coupled with differential privacy mechanisms. This solution is tailored to the unique characteristics of smart healthcare data, such as its volume, and variety of data types and distributions. By enriching and augmenting the input data, our approach generates realistic synthetic data samples that preserve the statistical properties of the original dataset. Furthermore, our proposed approach shows its capability to generate differentially private data samples, even when learning from a noisy distribution or adding noise to the learned distribution. To validate the effectiveness of our approach, we conduct evaluations on a real-world Fitbit dataset, demonstrating its ability to generate high-quality synthetic differentially private datasets.

Subsequently, we present a novel system for time series forecasting of health data streams using Spiking Neural Networks (SNNs), which are known for their energy-saving capabilities. The proposed system is designed to be more energy-efficient than the state-of-the-art LSTM-based counterpart. We conducted a comparative analysis between the SNN model and the LSTM-based model, revealing that the SNN model achieves a trade-off between accuracy and various efficiency metrics, including model size, parameter count, memory consumption, and training time. Moreover, our proposed system exhibits significantly lower energy consumption compared to traditional Artificial Neural Networks (ANNs), with an estimated energy consumption of 3.36 µJ. Furthermore, we integrate ε-differential privacy into our pipeline and perform an in-depth evaluation of its impact on the accuracy of both the baseline and clustered models.

We conclude from our aforementioned results that the SNN can be a great alternative to deepNets for edge applications such as radar applications and fitness trackers. We believe that this work will pave the way for the practical deployment of SNN-based edge applications in the future.

6.2 Future Work

Our future plans involve testing our gesture sensing and air-writing technology on larger and more complex datasets. Additionally, we aim to map out the processing pipeline, including other signal processing blocks such as coherent pulse integration, moving target indication filtering, target detection, range-Doppler image, angle of arrival, and micro-Doppler behavior, which will enable us to classify micro-Doppler gestures. This will reduce significantly energy consumption and overall latency. We also intend to investigate non-parametric Fourier transforms to reduce energy.

Currently, our air-writing system uses a metal marker, but in the future, we hope to switch to a finger and mount the radar on a physical device such as a laptop. Furthermore, we plan to expand our dataset to include other special characters like diacritical marks.

Regarding time-series forecasting, we can extend our smart healthcare data generation solution to other domains such as smart homes, creating realistic synthetic datasets that protect highly sensitive user data. Moreover, to generalize our approach, we can utilize the spiking LMU method on other smart home datasets.

Bibliography

1. I. Goodfellow, J. Pouget-Abadie, M. Mirza, et al., "Generative adversarial networks", *Advances in Neural Information Processing Systems*, vol. 3, 2014.
2. M. I. Skolnik, *Introduction to radar systems /2nd edition/*. 1980.
3. R. M. Page, *The Origin of Radar*. Doubleday & Company, 1962, pp. 111–158.
4. W.-K. Chen, *The electrical engineering handbook*. Elsevier Academic Press, 2005, pp. 111–158.
5. J. Rovňaková and D. Kocur, "Weak signal enhancement in radar signal processing", in *20th International Conference Radioelektronika 2010*, 2010, pp. 1–4.
6. A. Stove, "Linear fmcw radar techniques", English, *IEE Proceedings F (Radar and Signal Processing)*, vol. 139, 343–350(7), 5 1992.
7. S. Rao, *Introduction to mmwave sensing: Fmcw radars*, [Online] https://www.ti.com/content/dam/videos/external-videos/2/3816841626001/5415528961001.mp4/subassets/mmwaveSensing-FMCW-offlineviewing_0.pdf, Accessed: August 23, 2023.
8. W. S. McCulloch and W. Pitts, "A logical calculus of the ideas immanent in nervous activity", in *Neurocomputing: Foundations of Research*. Cambridge, MA, USA: MIT Press, 1988, pp. 15–27.
9. F. Rosenblatt, *The Perceptron, a Perceiving and Recognizing Automaton Project Para* (Report: Cornell Aeronautical Laboratory). Cornell Aeronautical Laboratory, 1957.
10. R. Bracewell, *The Fourier Transform and Its Applications* (Circuits and systems). McGraw Hill, 2000.
11. F. Murtagh, "Multilayer perceptrons for classification and regression", *Neurocomputing*, vol. 2, no. 5–6, pp. 183–197, 1991.
12. C. C. Aggarwal, *Neural Networks and Deep Learning: A Textbook*. Springer, 2018.
13. D. H. Hubel and T. N. Wiesel, "Receptive fields of single neurones in the cat's striate cortex", *The Journal of Physiology*, vol. 148, no. 3, pp. 574–591, 1959. eprint: https://physoc.onlinelibrary.wiley.com/doi/pdf/10.1113/jphysiol.1959.sp006308.
14. Y. L. Cun, B. Boser, J. S. Denker, et al., "Advances in neural information processing systems 2", in, D. S. Touretzky, Ed., San Francisco, CA, USA: Morgan Kaufmann Publishers Inc., 1990, ch. Handwritten Digit Recognition with a Back-propagation Network, pp. 396–404.
15. A. Khan, A. Sohail, U. Zahoora, and A. S. Qureshi, "A survey of the recent architectures of deep convolutional neural networks", *Artificial Intelligence Review*, vol. 53, pp. 5455–5516, 2020.

16. G. E. Dahl, T. N. Sainath, and G. E. Hinton, "Improving deep neural networks for lvcsr using rectified linear units and dropout", in *2013 IEEE International Conference on Acoustics, Speech and Signal Processing*, 2013, pp. 8609–8613.
17. D. E. Rumelhart, G. E. Hinton, and R. J. Williams, "Learning representations by backpropagating errors", *Nature*, vol. 323, no. 6088, pp. 533–536, 1986.
18. S. Hochreiter and J. Schmidhuber, "Long short-term memory", *Neural Computation*, vol. 9, no. 8, pp. 1735–1780, 1997.
19. J. Chung, C. Gulcehre, K. Cho, and Y. Bengio, "Empirical evaluation of gated recurrent neural networks on sequence modeling", *arXiv preprint* arXiv:1412.3555, 2014.
20. A. Santra, "Cognitive architectures, processing and learning algorithms for intelligent radar solutions", doctoralthesis, Friedrich-Alexander-Universität Erlangen-Nürnberg (FAU), 2022.
21. C. Lea, M. D. Flynn, R. Vidal, A. Reiter, and G. D. Hager, "Temporal convolutional networks for action segmentation and detection", in *proceedings of the IEEE Conference on Computer Vision and Pattern Recognition*, 2017, pp. 156–165.
22. A. van den Oord, S. Dieleman, H. Zen, *et al.*, *Wavenet: A generative model for raw audio*, 2016. arXiv: 1609.03499 [cs.SD].
23. T. S. Kim and A. Reiter, "Interpretable 3d human action analysis with temporal convolutional networks", 2017, pp. 1623–1631.
24. N. Kasabov, K. Dhoble, N. Nuntalid, and G. Indiveri, "Dynamic evolving spiking neural networks for on-line spatio- and spectro-temporal pattern recognition", *Neural Networks*, vol. 41, pp. 188–201, 2013, Special Issue on Autonomous Learning.
25. R. D. Hjelm, A. P. Jacob, T. Che, A. Trischler, K. Cho, and Y. Bengio, "Boundary-seeking generative adversarial networks", *arXiv preprint* arXiv:1702.08431, 2017.
26. A. Valentian, S. Narduzzi, M. Arsalan, *et al.*, "Tools and methodologies for training, profiling, and mapping a neural network on a hardware target", in *Intelligent Edge-Embedded Technologies for Digitising Industry*, O. Vermesan and M. D. Nava, Eds., River Publishers Series in Communications and Networking, 2022.
27. S. B. Furber, F. Galluppi, S. Temple, and L. A. Plana, "The spinnaker project", *Proceedings of the IEEE*, vol. 102, no. 5, pp. 652–665, 2014.
28. F. Akopyan, J. Sawada, A. Cassidy, *et al.*, "Truenorth: Design and tool flow of a 65 mw 1 million neuron programmable neurosynaptic chip", *IEEE Transactions on Computer-Aided Design of Integrated Circuits and Systems*, vol. 34, no. 10, pp. 1537–1557, 2015.
29. M. Davies, N. Srinivasa, T.-H. Lin, *et al.*, "Loihi: A neuromorphic manycore processor with on-chip learning", *IEEE Micro*, vol. 38, no. 1, pp. 82–99, 2018.
30. G. Orchard, E. P. Frady, D. B. D. Rubin, *et al.*, *Efficient neuromorphic signal processing with loihi 2*, 2021. arXiv: 2111.03746 [cs.ET].
31. J. D. Nunes, M. Carvalho, D. Carneiro, and J. S. Cardoso, "Spiking neural networks: A survey", *IEEE Access*, vol. 10, pp. 60 738–60 764, 2022.
32. A. L. Hodgkin and A. F. Huxley, "A quantitative description of membrane current and its application to conduction and excitation in nerve", *The Journal of Physiology*, vol. 117, no. 4, pp. 500–544, 1952.
33. W. Gerstner and W. M. Kistler, "The generalization of the leaky integrate-and-fire model", *Biological Cybernetics*, vol. 87, no. 5–6, pp. 469–482, 2002.
34. S. Dutta *et al.*, "Leaky integrate and fire neuron by charge-discharge dynamics in floating-body mosfet", *Sci. Rep.*, vol. 7, 2017.

35. D. Auge, J. Hille, E. Mueller, and A. Knoll, "A survey of encoding techniques for signal processing in spiking neural networks", English, *Neural Processing Letters*, vol. 53, no. 6, pp. 4693–4710, 2021.
36. T. Bekolay *et al.*, "Nengo: A python tool for building large-scale functional brain models", *Front. Neuroinform.*, vol. 7, p. 48, 2014.
37. J. A. K. Ranjan, T. Sigamani, and J. Barnabas, "A novel and efficient classifier using spiking neural network", *The Journal of Supercomputing*, vol. 76, no. 9, pp. 6545–6560, 2020.
38. D. Rasmussen, "Nengodl: Combining deep learning and neuromorphic modelling methods", *Neuroinformatics*, vol. 17, no. 4, pp. 611–628, 2019.
39. V. Senft *et al.*, "Reduction of dopamine in basal ganglia and its effects on syllable sequencing in speech: A computer simulation study", *Basal Ganglia*, vol. 6, no. 1, pp. 7–17, 2016.
40. B. J. Kröger *et al.*, "Modeling speech production using the neural engineering framework", in *5th CogInfoCom*, IEEE, 2014, pp. 203–208.
41. K. E. Friedl *et al.*, "Human-inspired neurorobotic system for classifying surface textures by touch", *IEEE Robot. Autom. Lett.*, vol. 1, no. 1, pp. 516–523, 2016.
42. J. Knight *et al.*, "Efficient spinnaker simulation of a heteroassociative memory using the neural engineering framework", in *IJCNN*, IEEE, 2016, pp. 5210–5217.
43. A. R. Voelker and C. Eliasmith, "Improving spiking dynamical networks: Accurate delays, higher-order synapses, and time cells", *Neural Comput.*, vol. 30, no. 3, pp. 569–609, 2018.
44. A. Voelker *et al.*, "Legendre memory units: Continuous-time representation in recurrent neural networks", vol. 32, 2019.
45. N. Chilkuri *et al.*, "Language modeling using lmus: 10x better data efficiency or improved scaling compared to transformers", *arXiv preprint* arXiv:2110.02402, 2021.
46. H. B. McMahan *et al.*, "Federated learning of deep networks using model averaging", *CoRR*, 2016.
47. B. McMahan *et al.*, "Communication-efficient learning of deep networks from decentralized data", in *AISTATS*, PMLR, 2017, pp. 1273–1282.
48. S. Imtiaz, "Privacy preserving behaviour learning for the iot ecosystem", PhD thesis, KTH Royal Institute of Technology, Stockholm, 2021.
49. Q. Geng and P. Viswanath, "The optimal noise-adding mechanism in differential privacy", *IEEE Transactions on Information Theory*, vol. 62, no. 2, pp. 925–951, 2015.
50. C. Dwork, A. Roth, *et al.*, "The algorithmic foundations of differential privacy.", *FnT-TCS*, vol. 9, no. 3-4, pp. 211–407, 2014.
51. P. Molchanov *et al.*, "Multi-sensor system for driver's hand-gesture recognition", in *11th IEEE FG*, vol. 1, 2015, pp. 1–8.
52. X. Zabulis *et al.*, "Vision-based hand gesture recognition for human-computer interaction", in *The Universal Access Handbook*, 2009.
53. X. Ma and J. Peng, "Kinect sensor-based long-distance hand gesture recognition and fingertip detection with depth information", *J. Sens.*, pp. 1–9, 2018.
54. A. K. Malima *et al.*, "A fast algorithm for vision-based hand gesture recognition for robot control", in *14th IEEE SIU*, 2006, pp. 1–4.
55. D.-S. Tran *et al.*, "Real-time hand gesture spotting and recognition using RGB-D camera and 3D convolutional neural network", *Appl. Sci.*, vol. 10, p. 722, 2020.

56. R. Ramalingame et al., "Wearable smart band for american sign language recognition with polymer carbon nanocomposite-based pressure sensors", *IEEE Sens. Lett.*, vol. 5, no. 6, pp. 1–4, 2021.
57. D. Jiang et al., "Hand gesture recognition using three-dimensional electrical impedance tomography", *IEEE Trans. Circuits Syst. II Express Briefs*, vol. 67, no. 9, pp. 1554–1558, 2020.
58. S.-W. Byun and S.-P. Lee, "Implementation of hand gesture recognition device applicable to smart watch based on flexible epidermal tactile sensor array", *Micromachines*, vol. 10, p. 692, 2019.
59. M. Georgi et al., "Recognizing hand and finger gestures with imu based motion and emg based muscle activity sensing", in *Proc. of the Int. Joint Conf. on BIOSTEC*, vol. 4, 2015, pp. 99–108.
60. A. Ferrone et al., "Wearable band for hand gesture recognition based on strain sensors", in *6th IEEE BioRob*, 2016, pp. 1319–1322.
61. F.-K. Wang et al., "Gesture sensing using retransmitted wireless communication signals based on doppler radar technology", *IEEE TMTT*, vol. 63, no. 12, pp. 4592–4602, 2015.
62. T. Fan et al., "Wireless hand gesture recognition based on continuous-wave doppler radar sensors", *IEEE TMTT*, vol. 64, no. 11, pp. 4012–4020, 2016.
63. Y. Zhang et al., "Hand gesture recognition for smart devices by classifying deterministic doppler signals", *IEEE TMTT*, vol. 69, no. 1, pp. 365–377, 2021.
64. V. Lammert et al., "A 122 ghz ism-band fmcw radar transceiver", in *13th GeMiC*, 2020, pp. 96–99.
65. V. Issakov et al., "A highly integrated d-band multi-channel transceiver chip for radar applications", in *IEEE BCICTS*, 2019.
66. J. Rimmelspacher et al., "Low power low phase noise 60 GHz multichannel transceiver in 28 nm CMOS for radar applications", in *IEEE RFIC*, 2020, pp. 19–22.
67. A. Bilato et al., "A multichannel d-band radar receiver with optimized lo distribution", *IEEE SSCL*, vol. 4, pp. 141–144, 2021.
68. E. Aguilar et al., "A fundamental-frequency 122 ghz radar transceiver with 5.3 dbm single-ended output power in a 130 nm sige technology", in *2020 IEEE/MTT-S IMS*, 2020, pp. 1215–1218.
69. E. Aguilar et al., "Highly-integrated scalable d-band receiver front-end modules in a 130 nm sige technology for imaging and radar applications", in *2020 GeMiC*, 2020, pp. 68–71.
70. E. Aguilar et al., "A 130 ghz fully-integrated fundamental-frequency d-band transmitter module with > 4 dbm single-ended output power", *IEEE Trans. Circuits Syst. II Express Briefs*, vol. 67, no. 5, pp. 906–910, 2020.
71. V. Issakov et al., "Highly-integrated low-power 60 ghz multichannel transceiver for radar applications in 28 nm cmos", in *2019 IEEE MTT-S IMS*, 2019, pp. 650–653.
72. M. Pavlo et al., "Online detection and classification of dynamic hand gestures with recurrent 3d convolutional neural networks", in *IEEE CVPR*, 2016.
73. S. Rautaray et al., "Vision based hand gesture recognition for human computer interaction: A survey", *Artif. Intell. Rev.*, 2015.
74. J. Lien et al., "Soli: Ubiquitous gesture sensing with millimeter wave radar", *ACM Trans. Graph.*, vol. 35, no. 4, 2016.

75. S. Hazra and A. Santra, "Robust gesture recognition using millimetric-wave radar system", *IEEE Sens. Lett.*, vol. 2, no. 4, pp. 1–4, 2018.
76. A. Santra and S. Hazra, *Deep learning applications of short range radars*, 2020.
77. Z. Zhang et al., "Latern: Dynamic continuous hand gesture recognition using fmcw radar sensor", *IEEE Sens. J.*, 2018.
78. Y. Sun et al., "Multi-feature encoder for radar-based gesture recognition", in *IEEE RADAR*, 2020, pp. 351–356.
79. N. Kern et al., "Robust doppler-based gesture recognition with incoherent automotive radar sensor networks", *IEEE Sens. Lett.*, vol. 4, no. 11, pp. 1–4, 2020.
80. M. Altmann et al., "Multi-modal cross learning for an fmcw radar assisted by thermal and rgb cameras to monitor gestures and cooking processes", *IEEE Access*, vol. 9, pp. 22 295–22 303, 2021.
81. K. Ishak et al., "Human gesture classification for autonomous driving applications using radars", in *IEEE MTT-S ICMIM*, 2020, pp. 1–4.
82. M. Q. Nguyen et al., "Range-gating technology for millimeter-wave radar remote gesture control in iot applications", in *2018 IEEE MTT-S IWS*, 2018, pp. 1–4.
83. C. Gu et al., "Motion sensing using radar: Gesture interaction and beyond", *IEEE Microw. Mag.*, vol. 20, no. 8, pp. 44–57, 2019.
84. X. Cai et al., "One-shot radar-based gesture recognizer for fast prototyping", in *IEEE Sens. J.*, 2021, pp. 1–4.
85. E. Hayashi et al., "Radarnet: Efficient gesture recognition technique utilizing a miniature radar sensor", in *CHI '21*, 2021, pp. 1–14.
86. M. Scherer et al., "TinyRadarNN: Combining Spatial and Temporal Convolutional Neural Networks for Embedded Gesture Recognition With Short Range Radars", *IEEE Internet Things J.*, vol. 8, no. 13, pp. 10 336–10 346, 2021.
87. Y. Sun et al., "Real-time radar-based gesture detection and recognition built in an edge-computing platform", *IEEE Sens. J.*, vol. 20, no. 18, pp. 10 706–10 716, 2020.
88. Y. Ren et al., "Hand gesture recognition using 802.11ad mmwave sensor in the mobile device", in *IEEE WCNCW*, 2021, pp. 1–6.
89. A. Helen Victoria and G. Maragatham, "Gesture recognition of radar micro doppler signatures using separable convolutional neural networks", *Materials Today: Proceedings*, 2021.
90. M. G. Amin, Z. Zeng, and T. Shan, "Hand gesture recognition based on radar micro-doppler signature envelopes", in *2019 IEEE Radar Conference (RadarConf)*, 2019, pp. 1–6.
91. M. Ritchie and A. M. Jones, "Micro-doppler gesture recognition using doppler, time and range based features", in *2019 IEEE Radar Conference (RadarConf)*, 2019, pp. 1–6.
92. V. Chen, F. Li, S.-S. Ho, and H. Wechsler, "Micro-doppler effect in radar: Phenomenon, model, and simulation study", *IEEE Transactions on Aerospace and Electronic Systems*, vol. 42, no. 1, pp. 2–21, 2006.
93. V. Sze et al., "Efficient processing of deep neural networks: A tutorial and survey", *Proc. of the IEEE*, vol. 105, no. 12, pp. 2295–2329, 2017.
94. M. Ester, H. Kriegel, J. Sander, X. Xu, et al., "A density-based algorithm for discovering clusters in large spatial databases with noise.", in *KDD*, vol. 96, 1996, pp. 226–231.
95. W. Maass, "Networks of spiking neurons: The third generation of neural network models", *Neural Networks*, vol. 10, no. 9, pp. 1659–1671, 1997.

96. G. Indiveri and T. Horiuchi, "Frontiers in neuromorphic engineering", *Front. Neurosci.*, vol. 5, p. 118, 2011.
97. M. Pfeiffer and T. Pfeil, "Deep learning with spiking neurons: Opportunities and challenges", *Front. Neurosci.*, vol. 12, p. 774, 2018.
98. P. Panda *et al.*, "Toward scalable, efficient, and accurate deep spiking neural networks with backward residual connections, stochastic softmax, and hybridization", *Front. Neurosci*, vol. 14, p. 653, 2020.
99. D.-A. Nguyen *et al.*, "A review of algorithms and hardware implementations for spiking neural networks", *J. Low Power Electron. Appl.*, vol. 11, no. 2, 2021.
100. H. Shouval, S. Wang, and G. Wittenberg, "Spike timing dependent plasticity: A consequence of more fundamental learning rules", *Frontiers in Computational Neuroscience*, vol. 4, 2010.
101. A. Safa, I. Ocket, A. Bourdoux, H. Sahli, F. Catthoor, and G. Gielen, *A new look at spike-timing-dependent plasticity networks for spatio-temporal feature learning*, 2021.
102. H. M. Lehmann, J. Hille, C. Grassmann, and V. Issakov, "Leaky integrate-and-fire neuron with a refractory period mechanism for invariant spikes", in *2022 17th Conference on Ph.D Research in Microelectronics and Electronics (PRIME)*, 2022, pp. 365–368.
103. S. Trotta *et al.*, "2.3 soli: A tiny device for a new human machine interface", in *2021 IEEE ISSCC*, vol. 64, 2021, pp. 42–44.
104. E. Hunsberger and C. Eliasmith, "Training spiking deep networks for neuromorphic hardware", *arXiv preprint* arXiv:1611.05141, 2016.
105. N. Kasabov, *Time-Space, Spiking Neural Networks and Brain-Inspired Artificial Intelligence*. Springer Berlin Heidelberg, 2018.
106. V. V. Chudnikov *et al.*, "Doa estimation in radar sensors with colocated antennas", in *2020 SYNCHROINFO*, 2020, pp. 1–6.
107. M. Chmurski *et al.*, "Highly-optimized radar-based gesture recognition system with depthwise expansion module", *Sensors*, vol. 21, no. 21, 2021.
108. J. Stuijt *et al.*, "μbrain: An event-driven and fully synthesizable architecture for spiking neural networks", *Front. Neurosci.*, vol. 15, p. 538, 2021.
109. D. P. Kingma and J. Ba, *Adam: A method for stochastic optimization*, 2014.
110. R. M. Schmidt, F. Schneider, and P. Hennig, *Descending through a crowded valley—benchmarking deep learning optimizers*, 2020.
111. X. Glorot and Y. Bengio, "Understanding the difficulty of training deep feedforward neural networks", *Journal of Machine Learning Research—Proceedings Track*, vol. 9, pp. 249–256, 2010.
112. A. Safa, F. Corradi, L. Keuninckx, *et al.*, "Improving the accuracy of spiking neural networks for radar gesture recognition through preprocessing", *IEEE Transactions on Neural Networks and Learning Systems*, pp. 1–13, 2021.
113. P. Blouw, X. Choo, E. Hunsberger, and C. Eliasmith, "Benchmarking keyword spotting efficiency on neuromorphic hardware", *ArXiv*, vol. abs/1812.01739, 2019.
114. T. Obaid, H. Rashed, A. Abou-Elnour, M. Rehan, M. Hasan, and M. Tarique, "Zigbee technology and its application in wireless home automation systems: A survey", *International journal of Computer Networks Communications*, vol. 6, pp. 115–131, 2014.
115. B. Alwaely and C. Abhayaratne, "Graph spectral domain feature learning with application to in-air hand-drawn number and shape recognition", *IEEE Access*, vol. 7, pp. 159 661–159 673, 2019.

116. M. S. Alam, K.-C. Kwon, M. A. Alam, M. Y. Abbass, S. M. Imtiaz, and N. Kim, "Trajectory-based air-writing recognition using deep neural network and depth sensor", *Sensors*, vol. 20, no. 2, 2020.
117. P. Roy, P. Kumar, S. Patidar, and R. Saini, "3d word spotting using leap motion sensor", *Multimedia Tools and Applications*, vol. 80, 2021.
118. T.-Y. Pan, C.-H. Kuo, H.-T. Liu, and M.-C. Hu, "Handwriting trajectory reconstruction using low-cost imu", *IEEE Transactions on Emerging Topics in Computational Intelligence*, vol. 3, no. 3, pp. 261–270, 2019.
119. V. Chandel, S. Singhal, and A. Ghose, "Airite: Towards accurate & infrastructure-free 3-d tracking of smart devices", in *2020 IEEE International Conference on Pervasive Computing and Communications Workshops (PerCom Workshops)*, 2020, pp. 1–6.
120. T. Yanay and E. Shmueli, "Air-writing recognition using smart-bands", *Pervasive and Mobile Computing*, vol. 66, p. 101 183, 2020.
121. W. Wang, A. X. Liu, and K. Sun, "Device-free gesture tracking using acoustic signals", in *Proceedings of the 22nd Annual International Conference on Mobile Computing and Networking*, ser. MobiCom '16, New York City, New York: Association for Computing Machinery, 2016, pp. 82–94.
122. C. Zhang, G. Abowd, Q. Xue, *et al.*, "Soundtrak: Continuous 3d tracking of a finger using active acoustics", *Proceedings of the ACM on Interactive, Mobile, Wearable and Ubiquitous Technologies*, vol. 1, pp. 1–25, 2017.
123. S. Yun, Y.-C. Chen, H. Zheng, L. Qiu, and W. Mao, "Strata: Fine-grained acoustic-based device-free tracking", in *Proceedings of the 15th Annual International Conference on Mobile Systems, Applications, and Services*, ser. MobiSys '17, Niagara Falls, New York, USA: Association for Computing Machinery, 2017, pp. 15–28.
124. Z. Fu, J. Xu, Z. Zhu, A. X. Liu, and X. Sun, "Writing in the air with wifi signals for virtual reality devices", *IEEE Transactions on Mobile Computing*, vol. 18, no. 02, pp. 473–484, 2019.
125. Y. Fang, Y. Xu, H. Li, X. He, and L. Kang, "Writing in the air: Recognize letters using deep learning through wifi signals", in *2020 6th International Conference on Big Data Computing and Communications (BIGCOM)*, 2020, pp. 8–14.
126. L. Sun, S. Sen, D. Koutsonikolas, and K.-H. Kim, "Widraw: Enabling hands-free drawing in the air on commodity wifi devices", in *Proceedings of the 21st Annual International Conference on Mobile Computing and Networking*, ser. MobiCom '15, Paris, France: Association for Computing Machinery, 2015, pp. 77–89.
127. M. Arsalan and A. Santra, "Character recognition in air-writing based on network of radars for human-machine interface", *IEEE Sens. J.*, vol. PP, pp. 1–1, 2019.
128. S. D. Regani, C. Wu, B. Wang, M. Wu, and K. J. R. Liu, "Mmwrite: Passive handwriting tracking using a single millimeter-wave radio", *IEEE Internet of Things Journal*, vol. 8, no. 17, pp. 13 291–13 305, 2021.
129. M. Arsalan *et al.*, "Air-writing with sparse network of radars using spatio-temporal learning", in *25th ICPR*, 2021, pp. 8877–8884.
130. Arsalan, Muhammad *et al.*, "Radar trajectory-based air-writing recognition using temporal convolutional network", in *19th IEEE ICMLA*, 2020, pp. 1454–1459.
131. A. Norrdine, "An algebraic solution to the multilateration problem", in *IEEE IPIN*, 2012.

132. D. L. Donoho, "Compressed sensing", *IEEE Transactions on Information Theory*, vol. 52, no. 4, pp. 1289–1306, 2006.
133. K. Kavukcuoglu et al., "Learning convolutional feature hierarchies for visual recognition", in *Proceedings of the NIPS*, Vancouver, British Columbia, Canada, 2010, pp. 1090–1098.
134. Y.-H. Liu and X.-J. Wang, "Spike-frequency adaptation of a generalized leaky integrate-and-fire model neuron", *Journal of computational neuroscience*, vol. 10, no. 1, pp. 25–45, 2001.
135. R. Zenun Franco et al., "Popular nutrition-related mobile apps: A feature assessment", *JMIR mhealth and uhealth*, vol. 4, 2016.
136. P. Voigt and A. Von dem Bussche, "The EU General Data Protection Regulation (GDPR)", *A Practical Guide, 1st Ed., Cham: Springer International Publishing*, 2017.
137. C. Dwork et al., "Calibrating noise to sensitivity in private data analysis", in *TCC*, Springer, 2006, pp. 265–284.
138. H. B. McMahan et al., "A general approach to adding differential privacy to iterative training procedures", *arXiv preprint arXiv:1812.06210*, 2018.
139. Ú. Erlingsson et al., "Rappor: Randomized aggregatable privacy-preserving ordinal response", in *Proceedings of the 2014 ACM SIGSAC CCS*, 2014, pp. 1054–1067.
140. M. Fredrikson, S. Jha, and T. Ristenpart, "Model inversion attacks that exploit confidence information and basic countermeasures", in *Proceedings of the 22nd ACM SIGSAC Conference on Computer and Communications Security*, 2015, pp. 1322–1333.
141. M. Veale, R. Binns, and L. Edwards, "Algorithms that remember: Model inversion attacks and data protection law", *Philosophical Transactions of the Royal Society A: Mathematical, Physical and Engineering Sciences*, vol. 376, no. 2133, p. 20 180 083, 2018.
142. J. Walonoski, M. Kramer, J. Nichols, et al., "Synthea: An approach, method, and software mechanism for generating synthetic patients and the synthetic electronic health care record", *Journal of the American Medical Informatics Association*, vol. 25, no. 3, pp. 230–238, 2018.
143. H. Li, L. Xiong, L. Zhang, and X. Jiang, "DPSynthesizer: Differentially private data synthesizer for privacy preserving data sharing", *Proc. VLDB Endow.*, vol. 7, no. 13, pp. 1677–1680, 2014.
144. M. Young et al., "Beyond open vs. closed: Balancing individual privacy and public accountability in data sharing", in *Proceedings of the ACM FAT*, 2019, pp. 191–200.
145. M. K. Baowaly, C.-C. Lin, C.-L. Liu, and K.-T. Chen, "Synthesizing electronic health records using improved generative adversarial networks", *Journal of the American Medical Informatics Association*, vol. 26, no. 3, pp. 228–241, 2019.
146. H. Ping, J. Stoyanovich, and B. Howe, "Datasynthesizer: Privacy-preserving synthetic datasets", in *Proceedings of the 29th International Conference on Scientific and Statistical Database Management*, 2017, pp. 1–5.
147. S. Imtiaz, R. Sadre, and V. Vlassov, "On the case of privacy in the IoT ecosystem: A survey", in *International Conference on Internet of Things (iThings)*, IEEE, 2019, pp. 1015–1024.
148. "Fitbit charge 2 user manual". Accessed: August 23, 2023, (), [Online]. Available: https://help.fitbit.com/manuals/manual_charge_2_en_US.pdf.

149. "Nutritionix API". Accessed: August 23, 2023, [Online]. ( ), Available: https://developer.nutritionix.com/.
150. F. J. Massey Jr, "The Kolmogorov-Smirnov test for goodness of fit", *Journal of the American statistical Association*, vol. 46, no. 253, pp. 68–78, 1951.
151. X.-S. Yang, *Nature-inspired optimization algorithms*. Academic Press, 2020.
152. A. Voelker, I. Kajić, and C. Eliasmith, "Legendre memory units: Continuous-time representation in recurrent neural networks", in *Advances in Neural Information Processing Systems*, H. Wallach, H. Larochelle, A. Beygelzimer, F. d'Alché-Buc, E. Fox, and R. Garnett, Eds., vol. 32, Curran Associates, Inc., 2019.
153. S. Imtiaz et al., "Synthetic and private smart health care data generation using gans", in *2021 International Conference on Computer Communications and Networks (ICCCN)*, 2021, pp. 1–7.
154. J. Stuijt et al., "μBrain: An Event-Driven and Fully Synthesizable Architecture for Spiking Neural Networks", *Frontiers in Neuroscience*, vol. 15, p. 538, 2021.

Printed in the USA
CPSIA information can be obtained
at www.ICGtesting.com
CBHW020718100924
14325CB00001B/37

9 783658 453176